ROUTLEDGE LIBRARY EDITIONS: ENERGY RESOURCES

Volume 3

NUCLEAR POWER IN INDIA

T0203633

NUCLEAR POWER IN INDIA
A Comparative Analysis

DAVID HART

Routledge
Taylor & Francis Group

LONDON AND NEW YORK

First published in 1983 by Allen & Unwin

This edition first published in 2019
by Routledge
2 Park Square, Milton Park, Abingdon, Oxon OX14 4RN

and by Routledge
52 Vanderbilt Avenue, New York, NY 10017

Routledge is an imprint of the Taylor & Francis Group, an informa business

British Library Cataloguing in Publication Data
A catalogue record for this book is available from the British Library

ISBN: 978-0-367-23168-2 (Set)
ISBN: 978-0-429-27857-0 (Set) (ebk)
ISBN: 978-0-367-23096-8 (Volume 3) (hbk)
ISBN: 978-0-367-23099-9 (Volume 3) (pbk)
ISBN: 978-0-429-27830-3 (Volume 3) (ebk)

Publisher's Note
The publisher has gone to great lengths to ensure the quality of this reprint but points out that some imperfections in the original copies may be apparent.

Disclaimer
The publisher has made every effort to trace copyright holders and would welcome correspondence from those they have been unable to trace.

The Indian nuclear power programme, both the earliest in the Third World and also one of the most comprehensive, provides an important and instructive subject for a wide-ranging and detailed study. This book, particularly timely when the prospects and problems for Third World nuclear power are commanding increasing attention, examines the origins and rationale of the Indian programme in the context of energy resources and consumption. It traces the progress of its historical development and leads up to an evaluation of the current performance, in both technical and economic terms of both individual reactors and the programme as a whole. In addtion, the book reveals high radiation exposure to workers at Tarapur, and discusses India's nuclear explosion of 1974 and the possibilities for novel developments in nuclear power and other energy sources, such as coal, biogas, hydro and solar power.

The author then sets the Indian programme into the world picture by comparing developments in India with those of the Third World (including developments in China and South Africa) and discusses the overall prospects for the Third World. The author's survey of nuclear power world-wide is a succinct but extremely informative account.

The author concludes that the performance of the Indian nuclear programme has not demonstrated any clear merits for nuclear power in that country up to the year 2000. The technical performance and the economics of the individual reactors, and the high cost of nuclear research and development, all add up to make the wisdom of embarking on a nuclear programme appear very doubtful. The case in favour of going nuclear appears to be extremely marginal for most other Third World countries also. Meanwhile, most Third World countries would be better advised to diversify both their R and D and their energy sources as much as possible.

It is not necessary to have any background in nuclear energy to understand this work, since the fundamentals of nuclear technology are clearly explained. The book will appeal to readers with general interests in energy; in science, technology and Third World developments; and in nuclear proliferation in the Third World.

The author brings to his study a scientific background and an interest in both technological development in the Third World and energy policy in general. He researched into technology and development (particularly hydroelectric energy and the role of the aluminium industry in Ghana's development) at Edinburgh University and subsequently researched this book while a Senior Research Assistant at Imperial College, London. David Hart was appointed Lecturer in Energy in the School of Environmental Sciences, University of East Anglia, in 1980.

Nuclear power in India:

A comparative analysis

DAVID HART

London
GEORGE ALLEN & UNWIN
Boston   Sydney

**George Allen & Unwin (Publishers) Ltd,
40 Museum Street, London WC1A 1LU, UK**

George Allen & Unwin (Publishers) Ltd,
Park Lane, Hemel Hempstead, Herts HP2 4TE, UK

Allen & Unwin Inc.,
9 Winchester Terrace, Winchester, Mass 01890, USA

George Allen & Unwin Australia Pty Ltd,
8 Napier Street, North Sydney, NSW 2060, Australia

First published in 1983

British Library Cataloguing in Publication Data

Hart, David
   Nuclear power in India.
   1. Atomic Power—India
   I. Title
   621.48′0954   TK9103
   ISBN   0-04-338101-4

Library of Congress Cataloging in Publication Data

Hart, David.
   Nuclear power in India.
   Bibliography: p.
   Includes index.
   1. Atomic power—India.   I. Title.
   TK9103.H37   1983   333.79′24′0954   82-18183
   ISBN   0-04-338101-4

Set in 11 on 12 point Times by Preface Ltd, Salisbury, Wilts.,
and printed in Great Britain
by Biddles Ltd, Guildford, Surrey

Contents

Acknowledgements

I would like to thank all those who assisted me in my research in the UK and in India. It is, however, only possible to name a few of the relevant persons and organisations. Dr Nigel Lucas and Professor Walter Murgatroyd supported my work by appointing me Senior Research Assistant in the Energy Policy Section of the Mechanical Engineering Department, Imperial College, London. Dr Anil Date and Professor A. Jaganmohan provided me with academic facilities and accommodation at the Indian Institute of Technology, Bombay. The Department of Atomic Energy, India, permitted me access to various facilities, including the Bhabha Atomic Research Centre and the Tarapur reactors. The University of London Central Research Fund made it possible for me to travel to India and spend six months there gathering data. Meg Heppenstall translated several works from their original German. Ainsley Rutledge, Ruth Ballard, Anna Phelps, Sue Winston, Jill Newnham and Sylvia Davies typed two drafts of my final report. To these named, and all the others, many thanks.

David Hart
January 1982

Abbreviations

AEB	Atomic Energy Board (South Africa)
AEC	Atomic Energy Commission (India)
AECL	Atomic Energy of Canada Ltd
AEO	Atomic Energy Organisation (Iran)
ASEA-ATOM	Swedish Electric Co.'s nuclear power subsidiary
BARC	Bhabha Atomic Research Centre (Bombay)
BHEL	Bharat Heavy Electricals Ltd (India)
BWR	boiling water reactor
CANDU	Canadian deuterium uranium reactor
CCEN	Chilean Nuclear Energy Commission
CEA	Central Electricity Authority (India)
CEA	Atomic Energy Commission (Cuba)
CEN	Nuclear Energy Commission (Cuba)
CIRUS or CIR	Canada–India reactor
CNEA	National Atomic Energy Commission (Argentina)
CNEN	National Nuclear Energy Commission (Brazil)
CWPC	Central Water and Power Commission (India)
DAE	Department of Atomic Energy (India)
D_2O	heavy water
ESCOM	Electricity Supply Commission (South Africa)
FBTR	fast breeder test reactor (India)
GE	General Electric (USA)
GELPRA	consortium of French and Swiss firms for heavy water production
GNP	gross national product
GW	gigawatts
IAEA	International Atomic Energy Agency
IBRD	International Bank for Reconstruction and Development
ICRP	International Commission on Radiological Protection
KANUPP	Karachi Nuclear Power Plant
KWU	Kraftwerk Union

LWR	light water reactor
MAPS	Madras Atomic Power Station
MWe	megawatts (electric)
MWth	megawatts (thermal)
NAPS	Narora Atomic Power Station (India)
NFC	Nuclear Fuel Complex (India)
NPD	Nuclear Power Demonstration Reactor (Canada)
NPT	Non-Proliferation Treaty
NRX	National Research Experimental reactor (Canada)
OECD/NEA	Organisation for Economic Co-operation and Development/Nuclear Energy Agency
PNE	peaceful nuclear explosion
PWR	pressurised water reactor
RAPS	Rajasthan Atomic Power Station (India)
SIPRI	Stockholm International Peace Research Institute
SWU	separative work units
TAPS	Tarapur Atomic Power Station (India)
TPC	Taiwan Power Corporation
UKAEA	United Kingdom Atomic Energy Authority
UNESCO	United Nations Educational Scientific and Cultural Organisation
URENCO	Uranium Enrichment Company
USAEC	United States Atomic Energy Commission
USAID	United States Agency for International Development

Introduction

The rise in the price of oil during 1973–4 focused the attention of the world on the subject of energy. Oil is now reckoned, in real terms, to be about five times as expensive as it was in 1972. This has led to particularly severe problems for the oil-importing underdeveloped countries. Does their future development depend upon a move away from oil use, or rather upon locating their own supplies? If the former, what alternative energy sources are they to choose?

The issue of nuclear power in the Third World is one that conflates three separate, and individually vital, topics. First, that of energy supplies for the world's future – whether these will consist of fossil fuels, nuclear power, renewable energy sources, or some other. Second, that of how the Third World is to develop – whether through establishing the techniques and institutions of the rich world and nurturing them in their own climes, or through fostering their own individual approaches which, it is hoped, would better satisfy their needs. Third, that of nuclear disarmament and weapons proliferation, a topic which is currently of greatly renewed interest for a number of disparate reasons (one of which is India's nuclear explosion of 1974). Thus, nuclear power in the Third World is of crucial interest to many people, in the First and Second Worlds as well as in the Third.

All three of these topics are touched upon in this work, but particular emphasis is given to the problem of energy. Is energy, as many people believe, in short supply worldwide? There are several indications that this is not so. On average about two-thirds of all the oil in every reservoir is left in the ground, because it is generally reckoned uneconomic to pump it out rather than just let it come up under its own pressure. About 40% of all the natural gas found in conjunction with oil is flared – this percentage being generally higher in the Third World. Although oil exploration activity has much increased world wide since the 1973 price rise (by

1

around 50%), the amount of activity in the oil-importing underdeveloped countries has remained more or less static. The multinational oil companies have shown a rather lukewarm interest in Third World oil, and have been completely indifferent to natural gas in the Third World because it is not easily exportable.

So it would appear that there is not a shortage of energy *per se*, but a shortage of cheap and readily available energy. The term 'energy crisis' is something of a misnomer; rather, what we have passed through, and are still passing through, is a radical change in the political economy of energy.

As far as the Third World is concerned, further discoveries of oil and gas would be very welcome; oil has manifold uses, and gas could be used in power generation if a grid of gas pipelines to industry is not economically worthwhile. Argentina, Brazil, Chile, Pakistan and Turkey are all trying to interest the oil companies in carrying out exploratory work. There are grounds for some optimism in the potential outcome of such work.

Those countries with substantial resources of coal (e.g. Brazil, India and Turkey) may well consider seriously the techniques of coal liquefaction and gasification as an alternative way of obtaining the 'free-flowing' fossil fuels. Such techniques are not well developed at present, but could well offer a much bigger return in the long run than investment in nuclear power.

This study of nuclear power in the Third World has lain particular emphasis on India's programme. There are two good reasons for this: first, India has one of the largest nuclear power programmes in the Third World and places a good deal of emphasis on self-reliance; second, India was the first Third World country to operate a commercial nuclear power station. Through a detailed analysis of the Indian programme, it has been possible to point out some of the problems and prospects for the rest of the Third World. This in turn could have strong relevance for the nuclear industries of the rich countries, for at present the lack of demand for nuclear power in the developed countries has meant that their nuclear industries have been forced to seek sales abroad in order to stay in business.

The presentation of topics is as follows. Chapter 1 gives an introduction to the technology of nuclear power and outlines some of its background history. Chapter 2 surveys India's production and consumption of energy and electricity. Chapter 3 reviews the history of India's nuclear programme (with an appendix on the nuclear explosion of 1974) and describes what state that programme has currently reached. Chapter 4 is a detailed assessment of the performance of nuclear power in India. Chapter 5 assesses the prospects for fast breeder reactors and for the use of thorium within India, and also gives a rapid summary of some of the alternative sources of energy open to India. Chapter 6 draws together some general considerations from the whole work and surveys the latest situation with regard to nuclear power in a number of key Third World countries, including all those which currently have nuclear power, and most of those which are likely to acquire it within the next few years.

A note on currencies: for the section on India I use rupees, while for the other sections I generally use US dollars. A table of exchange rates between the two is given:

Table *Exchange rates between US dollars and Indian rupees*

Year	Rupees per dollar
1955–65	4.8
1966–70	7.5
1971	7.2
1972	8.0
1973	8.1
1974	8.1
1975	8.9
1976	8.9
1977	8.2
1978	8.2
1979	8.2

CHAPTER 1

Essentials of Nuclear Power

Technical considerations

Fission and moderation
The basis of nuclear power is the fission process. This fission process is the splitting of an atom's nucleus, releasing energy in the form of heat, which can be converted, through a steam turbine and generator, into electricity. The only naturally occurring fissionable material is uranium 235 (^{235}U). This is a particular isotope of uranium having 235 neutrons and protons in its nucleus. Uranium found in the Earth's crust contains about one part of ^{235}U to 140 parts of ^{238}U. The only other naturally occurring uranium isotope is ^{234}U, which is present in such small quantities as to be negligible. Nuclear reactors can use as fuel either natural uranium with this ratio of ^{235}U to ^{238}U, or enriched uranium in which the proportion of ^{235}U has been increased.

The agent used to split the ^{235}U nucleus is a neutron, and since two or more neutrons are produced by each fission event it is possible to control the process such that one fission leads to another in a chain reaction. The neutrons produced as a consequence of fission have a substantial amount of energy because they are moving at high speed. Now ^{235}U captures fast neutrons (and is fissioned by them) about six times more readily than ^{238}U. But since in a natural uranium reactor there are about 140 times as many ^{238}U nuclei as there are ^{235}U nuclei, a chain reaction cannot be sustained with fast neutrons. However, slow neutrons ('thermal neutrons') are captured by ^{235}U about 200 times more strongly than by ^{238}U, therefore a chain reaction is

4

possible if the fission neutrons are slowed down or 'moderated'.

To slow the neutrons down quickly they must be made to collide with an element of low mass (the moderator), to which they will transfer a large part of their momentum. Only two substances appear to be suitable moderators for natural uranium reactors. These are heavy water, and carbon in the form of graphite. (Heavy water is just like ordinary water except that the hydrogen contained in ordinary water has been replaced by deuterium, an isotope of hydrogen. The hydrogen nucleus contains one proton, whereas the deuterium nucleus contains one neutron and one proton.) All other materials are either physically unsuitable or far too expensive.

If enriched fuel is used in a reactor (i.e. the ^{235}U content is raised to about 2–3%), the design constraints are considerably relaxed in comparison with those for a natural uranium reactor. The size of the fuel assembly can be smaller since one can afford to lose more neutrons from the system (this occurs essentially as a function of the ratio of surface area to volume). Also, neutron absorption by the moderator does not have to be minimised, so light water can be used instead of heavy water or graphite.

Reactor types
A majority of the world's operating reactors are based on enriched fuel. They are known as light water reactors (LWRs) since they use light water as coolant and as moderator. These reactors are essentially of American origin. LWRs may be divided into two types:

(a) Boiling water reactors (BWRs), in which the cooling water is allowed to boil inside the reactor vessel at a temperature of about 290 °C and a pressure of 70 atmospheres. The steam is then fed directly to the turbines and recirculated to the reactor. BWRs are made by the American firm General Electric. (Figure 1.1 gives a schematic illustration of a BWR.)

(b) Pressurised water reactors (PWRs), in which the pressure inside the reactor vessel is kept high (about 150 atmospheres) so as to prevent the cooling water

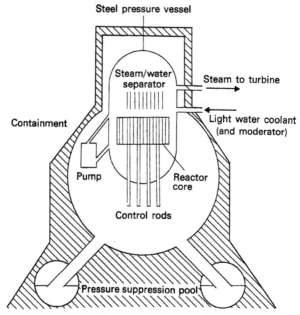

Figure 1.1 Schematic diagram of a boiling water reactor. The pressure suppression pool contains water which condenses any steam escaping from the pressure vessel and thereby reduces the resulting pressure inside the containment. The control rods are made of a neutron-absorbing material and are inserted into the core to shut it down or to reduce its reactivity.

from boiling at temperatures of up to 350 °C. This water is fed out of the reactor vessel (but still under pressure) to a steam generator where it passes through thousands of tubes immersed in water at a much lower pressure. This secondary cooling water boils and produces steam for the turbines. PWRs are made primarily by Westinghouse, but also by Combustion Engineering and Babcock and Wilcox (all of them American firms).

Of the kinds of reactor based on natural uranium the CANDU (CANadian Deuterium Uranium reactor) is one of the foremost. It uses heavy water as coolant and as moderator. The cooling system operates at a pressure of about 85 atmospheres and a temperature of 300 °C.

CANDUs are one of the most economical reactors in terms of fuel use. For a given energy output they use about 25–30% less natural uranium feed than do LWRs, and about half as much feed as the British style Magnox reactor.

The design of the CANDU is rather different to that of LWRs. Whereas the latter use one large steel pressure vessel containing the reactor core and coolant, the CANDU uses many pressure tubes in which lie individual fuel rods cooled by heavy water flowing through the length of the tube. The CANDU design is said to make fabrication a more routine, mass production process. (Figure 1.2 gives a schematic illustration of a CANDU reactor.)

One consequence of the different design philosphies is that CANDUs can be refuelled while constantly supplying

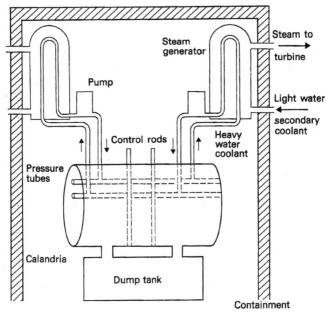

Figure 1.2 Schematic diagram of a CANDU reactor. The calandria is a cylindrical steel vessel which holds the heavy water moderator through which run the pressure tubes containing the fuel rods. The dump tank is where the heavy water moderator is transferred when it is desired to shut down the reactor. At the two ends of the calandria are fuelling machines which remove spent fuel and replace it with fresh fuel.

power. LWRs have to be shut down, usually on an annual basis, in order to change about a third of the fuel.

A further difference between the two types of reactor is that the CANDU tends to be more capital intensive, while the LWR costs more in terms of fuel. Very approximately, for a LWR, capital costs might account for 65% of total generating costs and fuel costs for 30%, whereas, for a CANDU, capital costs would make up 85% and fuel costs only 10%. In both cases the remaining 5% consists of operation and maintenance charges (Willrich, 1971a).

For both types of reactor, LWRs and CANDUs, it is generally considered that a minimum size for economic operation is around 500–600 MWe. Another common feature to these reactors is that they are less efficient in converting heat energy to electrical energy than are modern coal- or oil-fired power stations. Contrary to a common belief, the 'conventional' side of a nuclear power station (i.e. the turbine and generator) is not identical to that of a thermal power station. LWRs and CANDUs produce steam of a lower temperature and pressure ('wetter' steam) than modern thermal power stations. So turbines of larger dimensions are required which employ costly corrosion-resisting material for the blades, and have moisture-draining devices around the system.

All nuclear power reactors are intended to be used to supply continuous base-load to the grid system. There are two reasons for this. First, if a reactor is used in a load-following mode, the consequent changes in the reactor core tend to degrade the fuel and its cladding more quickly. Second, nuclear power is supposed to have high capital charges but low fuel costs, therefore it would make economic sense to go for maximum output in order to save higher fuel costs at other power stations and in order to pay off the capital debt more rapidly.

Nuclear fuel processing
The stages leading up to an assembled fuel rod ready to be placed in a reactor may be termed fuel processing. The steps involved are the mining and purifying of uranium; enrichment of the ^{235}U component if necessary; and the fabrication of the fuel rods.

The mining of uranium may be carried out in either open-cast or underground mines. The metal content of ore which is generally reckoned to be an economic proposition, and which is also the typical grade actually mined throughout the world, is around 0.1–0.25% uranium (i.e. ^{238}U and ^{235}U). This quality of ore compares rather unfavourably with other metallic ores extracted worldwide, which generally have a grade of 1% and upward (e.g. aluminium, 30%; iron, 30–70%; zinc, 3–8%). Silver and gold alone have lower grade ores and yet are still a commercial proposition. Other rare metals with low grade ores are generally mined in association with another metal to make the effort economically viable. Furthermore, the low grade of uranium ore is especially unfavourable when it is remembered that the fissionable part of the *purified* uranium is only 0.7%.

Of the non-communist world's reserves of uranium, 80% is located in four countries: the USA, Canada, South Africa and Australia. These countries are also currently the major producers of uranium.

The uranium ore originating from a mine is purified, usually in a nearby mill. Here the ore is crushed and chemically processed until only a mixture of uranium oxides remains, containing 85% uranium by weight.

If the uranium is destined for a natural uranium reactor, it is usually converted into uranium dioxide (which has a better behaviour at high temperatures than uranium metal) and then fabricated into fuel elements. However, if enriched fuel is to be used, the uranium must first be converted into a form which can be vapourised (uranium hexafluoride, UF_6). The enrichment process depends purely on the physical difference in mass between ^{235}U and ^{238}U. As the difference in mass between the two isotopes is fairly small, currently used enrichment processes require a large number of stages to produce a small percentage enrichment. Most enrichment plants to date have been based on the gaseous diffusion technique which uses the different rate of diffusion of the two isotopes through a membrane. Such plants are physically exceedingly large and use tremendous amounts of energy for pumping and cooling the UF_6 gas. The three enrichment plants in the USA consume a total of 6100 MWe of power when operating at full output (UN, 1972, vol. 9,

p. 36). For many years nearly 10% of the total electrical output of the USA was required to operate these plants while they were working at capacity to produce highly enriched material for weapons. A recently developed method of enrichment has been based on a gas centrifuge technique which tends to be a little less capital intensive and energy absorptive than the gaseous diffusion method, but uses even more technically sophisticated materials and equipment.

The fabrication stage of fuel processing involves the production of either uranium metal or refractory uranium dioxide, and their encasing in the form of long thin rods in a suitable material. This material must retain the radioactive fission products (the new atoms consisting of the fragments of ^{235}U) in the fuel, and yet not hinder the chain reaction process through neutron absorption. The material usually used these days is zircaloy. Zircaloy is an alloy of zirconium which is commonly used as a structural material in reactor cores because it has a low neutron capture cross-section and good resistance to corrosion even in water at high temperatures. It is, however, rather expensive. Fabrication is carried out in very clean conditions to avoid contaminating the fuel with neutron-absorbing foreign bodies.

Nuclear fuel reprocessing
After the fuel has been used in a nuclear reactor, there is still a considerable quantity of fissionable material left in it. The amount of ^{235}U in the fuel will, of course, have been reduced, but in enriched fuel it will only have been reduced to about the level pertaining in natural uranium, i.e. about 0.7%. Also, there will now be a new and artificial fissionable material present – plutonium. This is formed by ^{238}U capturing a neutron and transmuting into ^{239}Pu. Different reactors produce differing amounts of plutonium. The CANDU produces about twice as much plutonium per unit of electricity produced as does the LWR. (Although it must be noted that the concentration of plutonium in spent CANDU fuel is lower than in spent LWR fuel measured in, say, grams of plutonium per kilogram of fuel.)

The reason that the fuel is taken out before all the fissile material has been consumed is because it has become adulterated with fission products which interfere with the chain reaction process. Therefore, the aim of reprocessing is to separate out the fissile material from the waste.

Because the spent fuel is highly radioactive, it is stored for a time under water (which cools the fuel and also acts as a radiation shield). After about six months' storage its radioactivity has decreased sufficiently to make for easier handling. Nevertheless, the chemical reprocessing to separate the uranium and plutonium from the waste products must be carried out behind thick shielding. Also, since the separated plutonium can form a critical mass (i.e. a mass sufficient to support a spontaneous chain reaction, producing intense radiation and possibly an explosion), strict control must be maintained over quantities of this substance in vessels and pipelines.

Experience of reprocessing world wide has been limited mainly to uranium metal fuel which, unlike uranium dioxide, is never kept in a reactor for long and is therefore not as highly radioactive. The only power reactors of significance that use uranium metal fuel are the 'Magnox' style operated in Britain and France. The USA has built three plants for reprocessing LWR (i.e. UO_2) fuel, but none of them has been operated satisfactorily to date. Nuclear Fuel Services plant at West Valley, New York, cost $30 million to build and operated between 1966 and 1971 but was closed in 1972, ostensibly for modifications. There are no plans to re-open it, mainly because of severe problems in coping with radiation leaks and high-level waste products. General Electric began building a plant at Morris, Illinois, in 1968. After trying to operate it for two years and investing $64 million in it, they admitted, in 1974, that it would never work. Allied-General's plant at Barnwell, South Carolina, costing $360 million, underwent testing in 1978, but certainly acquired a doubtful future due to President Carter's opposition to reprocessing. The plant had not operated by late 1981. Allied-General wished to sell the plant to the government and continue as operating contractor, but the US government had been reluctant to buy.

Heavy water production
Heavy water reactors of all but the largest sizes require

Nuclear Power in India

about 1 tonne of heavy water to each electrical megawatt of capacity to cover moderating and cooling purposes. The most common method of producing heavy water is to extract it from ordinary water by what is known as the Girdler sulphide process. This process is based on the exchange of deuterium between hydrogen sulphide and water at different temperatures. The reaction, which occurs without a catalyst, is written as follows:

$$HDO + H_2S \leftrightharpoons H_2O + HDS$$

The process involves the use of tall towers, nearly 300 feet (90 metres) high, down which water flows across perforated trays. Hydrogen sulphide gas flows up through the trays and mixes with the water. The towers have hot and cold sections (130 °C and 30 °C respectively). In the cold sections, deuterium is transferred from gas to liquid. In the hot sections, it is transferred from liquid to gas. By repeating these exchanges several times the concentration of deuterium can be increased to about 30%, after which distillation is used to separate heavy water of about 99.8% purity or more from ordinary water.

Natural water contains about one part in 7000 of heavy water, and since only 20% of this can be economically removed (i.e. one part in 35 000) very large quantities of water must be processed. Even larger quantities of water are needed for cooling purposes. The need to handle such quantities makes the Girdler sulphide process energy intensive and capital intensive.

The Girdler sulphide process was developed in the USA in the 1950s, and was subsequently adopted in Canada in the 1960s. It is estimated that the USA and Canada have between them produced about 90% of the world's supplies almost entirely through the Girdler sulphide process. There are other processes available for heavy water production but none of them can be regarded as commercially proven.

Historical considerations

Nuclear power first became available in the mid 1950s. It was used to generate electricity in the UK, the USA, the

USSR and France. In the early 1960s these countries were followed by West Germany, Canada, Italy, Japan and Sweden. By 1970 Switzerland, East Germany, Holland, Spain, Belgium and India had nuclear power reactors. And, by 1980, Pakistan, Taiwan, Czechoslovakia, Argentina, Bulgaria, Finland and South Korea were all producing electricity by nuclear means, with Brazil about to start its first reactor in 1981.

Nuclear energy was, of course, first used in nuclear weapons at the end of World War II, and reactors were initially developed to produce plutonium for military purposes. For instance, even the first *commercial* nuclear power station in Britain, Calder Hall, which became operational in 1956, was primarily designed to produce plutonium and only secondarily to produce electricity. After World War II, in 1946, the US government passed the McMahon Act, which set up the US Atomic Energy Commission. The McMahon Act deemed it impossible to separate research on weapons from research on power generation at that stage in their development, and therefore labelled both topics as 'classified'. As a result co-operation on atomic energy research between the USA and other countries was not possible.

The military background to nuclear power was slowly shed throughout the 1950s, as it developed civil as well as military uses. In December 1953, US President Eisenhower made a speech to the UN General Assembly announcing an 'Atoms for Peace' plan. This plan was designed to share the peaceful benefits of American nuclear expertise with other countries, at the same time pre-empting the independent development of any military uses. Also, it was hoped that the plan would improve the American image abroad, which was still poor due to the memory of Hiroshima and Nagasaki (Nau *et al.*, 1976). Although these diplomatic and military objectives were paramount, the plan's aims opened up some commercial possibilities for the future. (At the time of Eisenhower's address nuclear power was an experimental rather than a commercial proposition.)

For several years after the 'Atoms for Peace' declaration, information on the costs of fuel, of reprocessing and plutonium, and heavy water, was still classified. Thus only rough estimates of fuel costs could be given to foreign customers. However, through such events as the series of conferences on the Peaceful Uses of Atomic Energy, held by the United Nations in 1955, 1958, 1964 and 1971, information has gradually become more and more available.

During the 1950s there was a boom in demand for uranium, almost entirely due to the military programme of the USA. This lapsed considerably in the 1960s, by which time stockpiles of weapons and uranium had been made. As a result, the price of uranium fell by more than 50% and a number of uranium mines were closed. Although nuclear power was becoming more widespread through the 1960s, it was not until the 1973 oil price rise that consumers decided to secure long term uranium supplies and the price of uranium started to climb again. The development from prototypes and military plutonium production reactors to fully fledged commercial reactors therefore occurred at a particularly favourable time (the 1960s) in terms of uranium prices.

Nuclear power and the USA
In February 1980, out of 405 000 MWe of nuclear power in operation or under construction world wide (including the Communist countries), the American domestic programme accounted for 182 000 MWe. Since there was also much American involvement in projects outside the USA, it is clear that that country had achieved a good measure of dominance over the nuclear power market. How had this been achieved?

There were some twenty firms in the USA engaged in nuclear reactor research and development in 1954. By 1960 this number had dropped to ten, and since 1966 only four American firms have remained in the reactor business, offering reactors on a commercial basis. The years 1963–6 saw General Electric and Westinghouse take over the nuclear power market almost completely. They appear to have done this through a series of a dozen low priced and fixed price turnkey contracts for reactors constructed in the USA and abroad. One of these contracts was the sale of the Tarapur twin unit BWR to India.

It has been estimated that General Electric and

Westinghouse incurred a loss of around $100 million each up to 1968 over these turnkey contracts (Allen and Melnik, 1975). *Fortune*, the business magazine, estimated in 1970 that GE had lost $200 million overall, having lost money on every turnkey plant it had built. And the *Wall Street Journal* in 1976 reported a former nuclear industry marketing executive as saying that the turnkey plants built (thirteen by GE and five by Westinghouse) had led to total losses for the two companies of between $800 million and $1 billion.

In 1966 General Electric announced that it would no longer offer reactors on a turnkey basis and Westinghouse did likewise. But the sale of these 'loss leaders' appears to have established General Electric and Westinghouse in a very strong position in the nuclear reactor industry. The other two firms, Babcock and Wilcox, and Combustion Engineering, have only a minority of the American domestic reactor market and have sold no reactors overseas. Whether the loss leader policy has been financially worthwhile seems doubtful. By the mid 1970s only Westinghouse claimed it was making a profit on its nuclear sales (*Business Week*, 1975). Subsequently this company has had to renege on its fixed price contracts to supply uranium fuel to utilities because an international uranium cartel dramatically raised the price of uranium supplies. A legal suit filed by Westinghouse against twenty-nine suppliers of uranium was due to go to trial in September 1981. Meanwhile, Westinghouse had been negotiating financial settlements with its utility customers; these settlements appeared to be losing Westinghouse hundreds of millions of dollars.

The first nuclear reactor in the world to be exported was an 11 MWe PWR sold by Westinghouse to Belgium in 1957. During 1957–73, of fifty-eight reactors exported around the world, the USA sold forty-six and Canada sold five. The American sales were split more or less evenly between Westinghouse and General Electric, and most of them went to developed countries. The proportion of American sales has been much reduced in more recent years, only amounting to sixteen orders out of forty-one world wide in the years 1974–7. This appears to be due, essentially, to a much developed nuclear capability and ability to export on the part of Europe.

Almost all American nuclear reactor exports to date have involved funding by the US Export-Import Bank. This is an offshoot of government that specialises in loans to foreign buyers of expensive American products such as aircraft and oil rigs. The Export-Import Bank sets interest rates about 2–5% lower than are commercially available, thus providing increased inducement to buy. Up to 1979 the Export-Import Bank had provided nearly $8 billion in loans and guarantees for nuclear exports from the USA. This amounted to about half of the sales cost of each of the sixty-seven reactors sold.

All the reactors exported from the USA use enriched fuel. Furthermore, the reactor suppliers have always supplied the fuel as well as the reactors. Enrichment has therefore been carried out at one of the three US government facilities, while fabrication has been carried out by either General Electric or Westinghouse. In fact the USA has had a virtual monopoly of the non-communist world's uranium enrichment facilities since World War II. Its three plants have had a capacity of 17 000 tonnes SWU (separative work units), recently expanded to 21 000 tonnes SWU. The only other plants in the non-communist world have been operated in Britain (400 tonnes SWU) and in France (300 tonnes SWU). The British plant, at Capenhurst, was opened for military use in 1952 and converted to civil use in the early 1960s. The French plant, at Pierrelatte, was opened in the early 1960s. (Other plants, of a much larger size, are now being set up in Britain and France, and also in Holland.)

The US Atomic Energy Act of 1954, and subsequent amendments, have required that the cost of enrichment services be recovered over a reasonable period of time. Recently proposed legislation has aimed to allow an additional percentage over these costs in the form of a profit. However, some expert observers have been of the opinion that enriched uranium may, in the past, have been sold at a price below real cost.

The upshot of these attractive prices for enriched uranium has been (a) a general lack of interest world wide in building uranium enrichment plants (thus contributing to a limitation of nuclear weapons proliferation); (b) a virtual monopoly over enriched fuel supplies by the U.S.A.; (c) a big success for American nuclear reactor exports (encouraged also by the early loss leader policy); (d) resultant influence on the part of the USA over nuclear developments around the world, by a degree of penetration and control of other countries' nuclear industries, through reactor and fuel sales.

However, in the long term the perception that the USA is using its dominance in civil nuclear power to further its foreign policy objective of nuclear non-proliferation (so maintaining its dominance of military nuclear technology) has led many countries to a wariness of being reliant on the USA. In the first instance this is leading to greater interest in nuclear supplies from Europe. Subsequently it could lead to a greater self-reliance through the acquisition of uranium enrichment, fuel fabrication and reprocessing plants. American propaganda labels this trend as one of weapons proliferation, but the countries concerned describe it as a drive for self-reliance in civil nuclear power.

Nuclear power and Canada

Similarly to that of the USA, the Canadian nuclear programme arose out of efforts during World War II to develop nuclear weapons. Canada was involved in a trilateral arrangement with the USA and the UK to develop nuclear know-how and materials, and when the war was over it was decided to put these to use in a civil programme. One of the substances the Canadians had studied during the war was heavy water, and they operated a small experimental production plant in the immediate postwar years. Having plenty of uranium ore, it was decided to base the Canadian programme on a once-through (i.e. no reprocessing) natural uranium heavy water reactor. (However, in recent years Canada has been seriously considering the use of plutonium extracted from spent fuel for 'advanced fuel cycles' in CANDU reactors involving fuels consisting of plutonium and natural uranium or plutonium and thorium.) The Canadian programme could be based on the CANDU, a more capital-expensive reactor than the American LWR, since Canadian power stations were financed by the public sector using interest rates of around 7%, while American power stations were privately financed using interest rates of about 12%. (The higher the interest rate the greater the economic penalty for capital intensity.)

Initially studies in Canada were carried out on research reactors only, but in April 1962 the Nuclear Power Demonstration Reactor, with an output of 22 MWe, went critical. Construction of the next stage in the programme, the Douglas Point 200 MWe reactor, began before this date, in January 1961. Douglas Point was expected to go into operation in 1964. Construction of the third stage in the programme, the Pickering nuclear power station, began in 1966, before Douglas Point went into commercial service in 1968. Douglas Point was, however, first operated in November 1966, so that *some* improvements were made at Pickering as a result of experience at Douglas Point. But a degree of haste in the Canadian programme has not left proper time for consideration and learning from past experience.

The performance of the Douglas Point reactor was generally regarded as disappointing. (See Chapter 4 for some of the technical details.) Although it was to have been a twin reactor station, the second unit was never built. Ontario Hydro, the state power utility, was committed by contract to offer to purchase Douglas Point from Atomic Energy of Canada Ltd when it had been demonstrated to be a safe and dependable source of power. The sale price was to be calculated to make the cost of electricity produced equal to that from a coal-fired power station. Despite these safe terms, Ontario Hydro has not purchased the reactor to date. The Canadians now tend to describe Douglas Point as a 'prototype' rather than a commercial proposition, whereas Pickering is regarded as a proven design.

In contrast to the American programme, the Canadian programme has been run essentially by a government corporation (Atomic Energy of Canada Ltd) with very little autonomy on the part of commercial enterprises. AECL does no manufacturing of its own, but contracts out work on nuclear equipment to a number of companies in North America. For instance, reactor cores are supplied by Canadian Vickers and Dominion Bridge, zircaloy pressure tubes by Chase Brass and Copper, boilers by Babcock and Wilcox Canada, and fuel by Canadian General Electric and

Westinghouse of Canada. Several of these companies have strong connections with the USA; 91% of Canadian General Electric is owned by General Electric of the U.S.A.

So although the CANDU reactor has been developed and marketed by the Canadians, it has not been entirely of Canadian production. For the Douglas Point reactor all the zircaloy components and heavy water were imported from the USA, and of the rest of the materials and equipment Canadian firms supplied 71% while the UK supplied 17% and the USA 12%. Until 1974 all zircaloy was imported from the USA. At that point in time the Canadian content represented about 80% of the capital cost of a CANDU. More recently, about 10% by value of the material and equipment for CANDUs was imported from the USA (Royal Commission on Electric Power Planning, 1978).

Like the USA, Canada uses special financing arrangements to encourage the export of nuclear reactors. The Export Development Corporation (like AECL, a Crown corporation wholly owned by the government) lent $129 million to Argentina to finance a CANDU power station in 1974, and in 1975 it organised a total loan of $330 million to South Korea for the same purpose.

Large loans at low interest rates do not appear to have been sufficient inducement to purchase CANDUs. During 1976 the Auditor General of Canada found that AECL had spent over $17 million on payments to sales agents for reactors sold abroad. More than $10 million of payments for sales made to South Korea and Argentina were inadequately documented. Six members of AECL's Board of Directors resigned in September 1976 and an investigation by the Parliamentary Standing Committee on Public Accounts was held between November 1976 and March 1978. This investigation covered the inadequate documentation of payments and also the losses expected on the contract with Argentina. The results of the investigation were inconclusive since the final recipients of the inadequately documented payments remained unidentified. However, the Committee expressed the opinion that AECL had, at best, been guilty of inadequate commercial judgement and, at worst, of bribery and corruption. The Company was also expected to make huge losses, largely as a result of its dealings with foreign

countries. The loss on the Argentine contract was put at over
$100 million.

The role of the IAEA

The International Atomic Energy Agency (IAEA) is
charged with two functions: to promote the peaceful use of
nuclear power around the world, and to regulate that use,
taking into account safety considerations, environmental
protection and safeguards against weapons production.
Throughout its history the IAEA has always spent more on
the promotion of nuclear power than on its regulation. In
particular, expenditure on implementing safeguards was low
during the 1960s, but grew throughout the 1970s, so that in
the latter part of the decade it made up one-fifth to a quarter
of the total IAEA budget.

CHAPTER 2

Energy and Electricity in India

Energy consumption

India has a fairly high consumption of commercial energy in comparison with gross national product when compared to other countries. For instance, China has a comparable level of energy consumption per capita but has a higher income per capita.

Estimates vary as to the precise quantities involved, but observers agree that total energy use in India has grown considerably over the last few decades. Commercial and non-commercial energy uses have grown, but the former has grown faster than the latter. (Commercial energy in India is mainly derived from fossil fuel and hydropower, whereas non-commercial energy is mainly supplied by firewood, cow dung and vegetable waste.) Consumption of commercial energy has doubled about every ten years, so that while in the 1950s non-commercial energy made up about two-thirds of total energy use, in the 1970s it made up less than one-half. Even so, in the early 1970s, the single most important energy source in India was still a non-commercial item namely, firewood.

Energy in India has been kept at a low price. Throughout the 1960s the wholesale prices of coal, electricity and oil all rose more slowly than prices as a whole. Indian coal is one of the cheapest in the world, although the black market price for coal in India is about twice the official price (Centre for Monitoring Indian Economy, 1979). This low price for energy, as well as a trend to 'modernisation', explains why the use of commercial energy has grown so rapidly.

21

Table 2.1 *Primary sources of commercial energy (in per cent)*

	1953/4	1960/1	1970/1
Coal	54	49	39
Oil	40	43	49
Hydro	6	8	11
Nuclear			1

Source: Henderson (1975, pp. 28 and 70)

The primary sources of commercial energy have (very approximately) been as shown in Table 2.1.

Energy resources

Proponents of nuclear power for India have never denied that the country has substantial sources of energy for its immediate needs. What they have said is, firstly, that indigenous conventional energy resources would never be sufficient to support a fully industrialised India in the long term; secondly, that exploitable energy is unevenly distributed so that nuclear power can be competitive now in certain areas of energy shortage, particularly the south and west of the country. Coal resources are said to be concentrated in the east, in the states of Bihar and West Bengal. Hydro potential is said to be mainly in the north in the foothills of the Himalayas.

These arguments have not been without contention. For instance, the Indian Government's Central Water and Power Commission stated in 1961 that the coal reserves together with the total hydro potential of India, when fully harnessed, appeared to be sufficient to enable the industrialisation of the country at the maximum rate that could be considered feasible during the next few decades. (Central Water and Power Commission, 1961, p. 2). This statement thus contradicts the long term and the short term arguments in favour of nuclear power.

The argument that energy resources in India are unevenly distributed ignores the extent to which a resource is not so much a fact as a product of human exploration and

Table 2.2   *Indian coal production and coal reserves*

Year	Coal production (million tonnes per year)	Coal reserves (billion tonnes)
1955	40	40
1965	70	80
1977	100	
1978		112

Until recently reserves were assessed to a depth of 600 metres and minimum seam thickness of 1.2 metres. Now, depths up to 1200 metres and seam thickness of 0.5 metres and above are used.

Sources: Henderson (1975, pp. 8 and 38); *Urja*, 10 June 1978; *Energy Management (India)*, July/September 1979, p. 239.

ingenuity. Coal and hydro resources have always been perceived as abundant in comparison with India's immediate needs, and therefore the incentive to explore for further resources has been almost non-existent.

For instance coal production and estimated reserves have been approximately as shown in Table 2.2.

Thus, despite a very low priority for exploration, reserves have kept pace with production so that the time to depletion (given constant consumption) has always been about 1000. years. Given this lack of pressure upon supplies, plus the low price of coal at the pithead, plus inexpensive rail transport charges and the high degree of concentration of industry around the coal fields in Bihar and West Bengal, it is hardly surprising that exploration has essentially been limited merely to extending known coal fields.

However, other areas are known to have supplies of coal. For instance the central states of Madhya Pradesh and Orissa are each thought to have over 30 billion tonnes of coal suitable for power generation; more than is estimated for Bihar and West Bengal combined. The National Council for Applied Economic Research stated in 1962 that the estimated resources of coal and lignite within the southern region were large, and represented nearly 150–200 times the level of anticipated demand within that region by 1975/6. Thus there is room for argument over any supposed 'maldistribution' of coal supplies.

Most of India's coal has a low sulphur content but a

Table 2.3   *Hydro potential in gigawatts at 60% load factor*

Region	CWPC survey	CEA reassessment
Northern	10.7	27.8
Eastern	2.7	7.5
Western	7.2	7.6
Southern	8.1	13.1
North-eastern	12.4	20.2
	41.1	76.2

Source: Ministry of Energy (1979, pp. 39–40).

moderately high ash content. In recent years there has been talk of establishing coal washeries to reduce the ash content before burning. This technique is in fairly widespread use elsewhere. Alternatively, fluidised bed combustion can cope with high ash fuel while also minimising pollution.

Similarly to coal, development of new hydro potential has received a low priority. The last major review of hydro potential was a study carried out by the erstwhile Central Water and Power Commission from 1953 to 1960. The Central Electricity Authority is only now carrying out a detailed updating of this work and a preliminary reassessment (Table 2.3) shows almost a doubling of total potential.

In 1978/9 the amount of hydro potential already developed was about 7.6 GW, while 5.2 GW were under development (Ministry of Energy, 1979). Thus about 17% of potential is either used or will be used in the fairly near future. This leaves quite a lot to spare.

Coal industry
The main problems of the Indian coal industry are not to do with insufficient reserves but with insufficient production to meet demand, poor labour conditions, and lack of transport. Henderson (1975, p. 104) describes a vicious circle whereby coal output is held down by power cuts, while power stations are restricted by shortage of coal. Comparison of figures for output per man-shift with developed countries' coal industries in 1970, shows, unsurprisingly, that India's output is rather low (see Table 2.4).

Table 2.4   *Coal output per man-shift*

Country	Coal output (tonnes per man-shift)
Australia	10.86
Belgium	1.60
Canada	9.03
Czechoslovakia	4.73
France	1.53
Holland	1.09
India	0.67
Japan	1.69
Poland	1.54
UK	2.14
USA	17.27
West Germany	2.21

Source: Parikh (1977, p. 5).

The reason, of course, is that automation and mechanisation are used to a much greater extent in the rich countries. Labour costs being low in India, there is very little incentive to mechanise on the ground of pure economics. However, mechanisation could well be seen as sensible, in order to (a) achieve greater reliability in production and (b) reduce deaths and accidents in the coal mines. Typical fatality rates in India and elsewhere, in the 1970s, were as shown in Table 2.5.

There is, therefore, plenty of scope for improvement on this aspect of the Indian coal industry.

The vast bulk of coal transported in India is carried by rail. Road and sea transport carry minimal quantities. In the

Table 2.5   *Deaths in the coal industry*

Country	Year	Fatalities per million tonnes produced
India	1973	3.63
	1977	2.37
UK	1973	0.56
USA	1973	0.21
EEC	1972	1.03

Sources: Urja, 10 June 1978; *National Coal Board Report 1973/4*; US Bureau of the Census (1980).

early 1960s rail transport was proving inadequate to carry sufficient coal to meet demand but from 1964/5 onward rail capacity was satisfactory. Unfortunately an era of wagon shortage began again in the early 1970s so that the mid 1970s saw a shortfall in coal supply once more. Since coal production increased slowly but steadily during this whole period, the basic problem would appear to lie with the railways rather than with unpredictability in coal production. The problem of shortage of railway wagons is probably exacerbated by the need to transfer load from gauge to gauge on some journeys. (There are three different track gauges in use in India.) Furthermore, even on a number of main lines Indian Railways make do with single track working. So although the Indian Railways system is one of the largest in the world, it would seem that there is a case for further investment and expansion.

Excluding the use of coking coal in the iron and steel industry, coal in India is used mainly for power generation and in railway engines. These uses make up one-third and one-quarter respectively of non-coking coal consumption.

The coal industry was nationalised in 1972/3. Since the oil industry and electricity supply are also essentially under state control, virtually all of India's energy industry is in the public sector.

Oil industry

The development of India's oil industry is interesting because it illustrates some of the general difficulties and prospects for energy in India.

When India gained independence in 1947 control of the oil industry remained largely in the hands of two British oil companies, Burmah and Shell. During the 1950s three major refineries were built in India, one by Burmah–Shell and two by American oil companies. The crude supply to these refineries was imported, and despite government attempts to persuade them differently, the oil companies remained reluctant to explore for oil supplies in India itself. The Indian government therefore set up its own exploration organisation, the Oil and Natural Gas Commission. During the late 1950s and 1960s, with Russian and Rumanian technical assistance, fairly substantial oil deposits were discovered. Exploration efforts have been continued to date and have been very successful. Only about 6% of the country's total oil requirements were produced indigenously before 1960, but indigenous production has met about one-third of consumption during the late 1960s and 1970s. Over the same period the government has followed a policy of increasing its participation in refining, through the building of public sector refineries and through taking over the private sector.

Prospects for further discoveries of oil appear to be good. A World Bank report (IBRD, 1979) has stated that an increase in exploration activity in India is warranted. Nearly half of India's landmass is made up of sedimentary areas (i.e. possible oil-bearing rock) and further promising areas lie offshore. By 1973 only 4% of the total sedimentary areas had been well studied, according to the report of the Fuel Policy Committee.

One difficulty in increasing exploration and output is the lack of a technological base for an oil industry in India. It is said that about 80% of equipment needed for onshore drilling has to be imported, and about 100% for offshore drilling (*Urja*, 28 October 1978). Thus, of the large investment expenditures needed to raise crude output, about one-half would have to be in foreign exchange.

Electricity supply industry

The electricity supply industry was essentially in private hands when India gained independence in 1947. The change to public control was gradual but steady, the private sector having 75% of installed capacity in 1950, less than 30% in 1966 and less than 10% in 1974.

Over the last twenty to thirty years slightly more than half of India's electricity has been produced by thermal generation (overwhelmingly coal-fired power stations) while slightly less than a half has been produced by hydropower. The rapid growth in the use of commercial energy has been paralleled in the electricity supply industry where demand continues to outpace supply. During the period 1955–65 the average annual rate of increase of power generation was around 13%, and during the period 1966–77 around 8%. A partial explanation for this rapid growth is the low price charged for electricity, especially that supplied to industry. Industry uses 65–70% of all electricity consumed in India and the rates charged to industry have been less than half those for domestic and commercial consumers (falling to well under one-third in the mid 1960s). In the UK, in contrast, the overall industrial rate is about three-quarters of the rate for domestic and commercial consumption.

Electricity prices are actually set by the state electricity boards, and typically the rates are set low to encourage the location of industry in that state. The wisdom of this policy is very doubtful since electricity costs make up at most around 3% of the total production costs in most industries. The price of electricity is therefore a marginal consideration in deciding where to build a new factory.

The state electricity boards do not charge a sufficient rate for electricity to make them financially autonomous, i.e. to provide capital for new power projects as well as covering costs. Instead they have to acquire funds from their respective state governments, in competition with other possible forms of spending. It has been recommended several times (for example, by the Venkataraman Committee in 1964 and by the Fuel Policy Committee, 1975, and Wagle and Rao, 1978) that the electricity boards be required to make their sales profitable enough to enable growth to occur through their own finances, but a positive step in this direction has not been made. The effects of increased charges for industrial electricity could be wholly beneficial. The electricity boards would be enabled to satisfy customer requirements through a combination of slackened demand and increased investment without imposing any power cuts (which have been a fact of life in India for many years). Enhanced reliability of supply would compensate industry for increased charges, while any marginal requirements for electricity which were dropped as a result of price increases would probably make industry slightly more labour-intensive, easing the problem of unemployment. Effects on the overall rate of industrialisation would be negligible. (See Venkataraman (1972, Ch. 5) for another argument that industrial tariffs in India are too low.)

As a result of their impoverished financial situation and the fact that nuclear power is financed by central government and not state governments, several state electricity boards have requested that nuclear power stations be set up within their boundaries so that they could avoid some capital expense. This appears to be a particularly artificial and unwise form of inducement to embark on a nuclear power programme.

The largest electrical generating sets being installed in India were of 140 MW size in the 1960s and of 200 MW size in the 1970s. By way of comparison the UK was installing 200 and 300 MW sets in the early 1960s, and 500 MW sets in the late 1960s. Transmission voltages in India have risen along with the size of generating set. Lines of 220 kV were installed in the early 1960s, and the first 400 kV line was commissioned in 1978.

Total installed generating capacity in India has grown steadily from around 2000 MWe in 1950 to over 14 000 MWe in 1969, and more than 25 000 MWe in 1978. In 1969 the peak demand was 8000 MWe while in 1975/6 it was 14 000 MWe. So peak demand seems generally to have been less than total capacity, but since there is no national grid, localised peak demands have frequently exceeded local supply.

By 1969, interconnections between states were still essentially inadequate for mutual power sharing. There were no viable regional grids, and certainly nothing like a national grid. During the Fourth Five Year Plan (1969/74) efforts were made to build a grid system. Inter-state and inter-regional transmission lines were treated as centrally sponsored schemes instead of being the responsibility of individual states. The first regional grid that could be said to be near adequately integrated went into operation in the Southern Region in 1972. It was estimated in 1972 that the cost of setting up something akin to a national grid would be about Rs 425 million. However, the operation of regional grids alone was estimated to reduce capacity requirement by the order of 100 MWe and more in each of the four main regions, simply through diversity of peak demand across each region.

So the minimum saving on capacity was reckoned to be

600 MWe, not including any reduction of spinning reserve capacity. An outlay of Rs 425 million for a saving of 600 MWe works out at around Rs 700 per kilowatt. Since at that time even the cheapest power stations (coal-fired) were costing twice that much (see page 80), it is clear that strengthening the grid makes very good economic sense.

It is, however, not clear that financial expenditure on the transmission and distribution system has been adequate over the years. A rough rule of thumb which is used internationally is that investment in generation and investment in transmission and distribution should be roughly equal. India spends about 50% more on generation than on transmission and distribution. The result is inadequate performance in terms of energy losses in the transmission and distribution system. These losses have been large and are rising year by year (Table 2.6).

According to Kelvin's law, transmission and distribution circuits are optimal when the annual cost of lost energy is equal to the annual fixed cost of the circuits. In India in 1974/5 the fixed cost chargeable to the transmission and distribution circuits was about Rs 1670 million, at an interest rate of 12%. The total value of energy lost calculated at 15 paise per kilowatt-hour came to Rs 2034 million. To make these figures equal one would need to invest a further Rs 3000 million in the transmission and distribution system.

Table 2.6   *Transmission and distribution losses*

Year	Loss (%)
1961/2	14.4
1965/6	14.0
1967/8	15.0
1968/9	16.7
1970/1	17.3
1975/6	19.4
1976/7	19.9

Sources: Fuel Policy Committee (1975, pp. 78 and 79); *Electrical India*, 15 June 1978, p. 29.

Table 2.7 *Bharat Heavy Electricals Ltd*

Plant	Equipment manufactured	Assistance
Tiruchirapalli	Boilers	Czechoslovak
Hyderabad	Steam turbine-generators	Czechoslovak
Hardwar	Hydro and steam turbine-generators	Soviet

Sources: Cilingiroglu (1969, pp. 34–5); *Economic and Political Weekly*, 11 March 1972, p. 570.

It appears fairly clear therefore that the transmission and distribution system has suffered from under-investment.

Manufacture of electric power supply equipment in India today is the responsibility of a public sector enterprise, Bharat Heavy Electricals Ltd (BHEL). The first of its four main plants, at Bhopal, went into production in the early 1960s. It produces both hydro and steam turbine-generator sets, and was commissioned with British assistance. Three other plants were built in the late 1960s. These were as shown in Table 2.7.

The capital investment in these three plants by 1971 (when the plants were in operation) was said to be about Rs 1455 million. Nowadays, BHEL employs around 56 000 people. Up to 1978 about 4400 MWe of thermal plant and 2300 MWe of hydro equipment made by BHEL had been installed and commissioned. The largest size of generating set that BHEL is capable of manufacturing these days is just over 200 MWe. In 1979 BHEL was seeking to pay Siemens, a West German firm, for know-how to develop larger generating sets.

Since the early 1970s, India has been theoretically capable of constructing almost all its new generation capacity indigenously. Generator sets have still been imported, nevertheless, mainly from the USSR, Czechoslovakia, Japan, the UK and the USA. In 1979 the Ministry of Energy compared the performance of BHEL sets with imported sets and found that the latter performed better. As a result the Ministry was not favourable to relying on indigenous sets alone.

CHAPTER 3

A History of India's
Nuclear Power Programme

The rationale

Although India is generally regarded as an underdeveloped country, it has never been very far behind the rich countries in the field of nuclear technology. An advisory body entitled the Atomic Energy Commission was established as early as 1948 in the Indian Ministry of Natural Resources and Scientific Research. In 1954 a Department of Atomic Energy with the full powers of a ministry was established with Jawaharlal Nehru (India's Prime Minister) as first Minister of Atomic Energy. In the same year the main atomic energy research establishment, now known as The Bhabha Atomic Research Centre (BARC), was set up on the outskirts of Bombay.

The man largely responsible for the development of India's nuclear power programme during its early years was Homi Bhabha, an eminent Indian physicist. He had obtained his university education at Cambridge in the late 1920s and 1930s and had associated with some of the great physicists working there in those days. Close links existed between the Bhabha family and the families of Tata (the industrialists) and Nehru. On 12 March 1944 Bhabha wrote to the Tata Trust to apply for money to set up an institute for fundamental scientific research. In his letter he expressed the hope that

'when nuclear energy has been successfully applied for power production in say a couple of decades from now, India will not have to look abroad for its experts but will find them ready at hand' (*Business India*, 4–17 September 1978, p. 20).

The Tata Institute of Fundamental Research was set up in 1945 with Bhabha as its director. He subsequently became Chairman of the AEC and Secretary of the DAE when these bodies were established, mainly at his request. From 1948 to 1957 membership of the AEC consisted of H. J. Bhabha, K. S. Krishnan and S. S. Bhatnagar, all of them scientists. In 1958 a full time member for finance was appointed, but in 1962 this post was altered to a part time position. Apparently, Bhabha did not want a full time member for finance interfering in the AEC's affairs. All the other members are appointed part time, so a full time member for finance would imply dual control of the AEC, except that the Chairman has the power to overrule.

As well as setting up India's nuclear power programme, Bhabha also initiated the Indian space programme through establishing the Indian National Committee for Space Research in 1962 under the chairmanship of Vikram Sarabhai. When Bhabha died in a plane crash on Mont Blanc in 1966, he was succeeded as Chairman of the AEC by Sarabhai. On Sarabhai's death in 1971, the post went to Homi Sethna, an engineer who had made his career in the DAE. Sethna is the current chairman of the AEC.

Bhabha and Nehru were obviously crucial figures in the establishment of a nuclear power programme in India. It is interesting to examine some of the opinions they expressed about nuclear power and about India's development.

Like Bhabha, Nehru received a substantial part of his education in England – at Harrow, Cambridge University and the Inner Temple. At Cambridge Nehru studied natural sciences (geology, chemistry and botany) and he remained devoted to science for the rest of his life, attending every annual session of the Indian Science Congress from 1947 to 1964, the year of his death. The following statements help us to understand his commitment to nuclear power:

'It is science alone that can solve the problem of hunger and poverty, of insanitation and illiteracy, of superstition and deadening custom and tradition, of vast resources running to waste, of a rich country inhabited by starving people. Who indeed can afford to ignore science today? At every turn we have to seek its aid. The future belongs to science and to those who make friends with science' (Morehouse, 1971, pp. 1–2).

'I do not see any way out of our vicious circle of poverty except by utilising the new sources of power which science has placed at our disposal' (Morehouse, 1971, p. 2).

'Power [i.e. electric power] is the most important thing for developing a country's resources. You may judge of a country's advance today merely by seeing how much power it produces or uses' (DAE, 1956, p. 2).

Bhabha expressed similar views:

'What the developed countries have and the underdeveloped lack, is modern science and an economy based on modern technology. The problem of developing the underdeveloped countries is, therefore, the problem of establishing modern science in them and transforming their economy to one based on modern science and technology' (Rahman and Sharma, 1974, p. 275).

'The key to all industrialisation is an adequate supply of electrical energy, and in its turn the electrical power consumption per head is as good an index as any of the standard of living' (Bhabha, 1964, p. 57).

'For the industrialisation of the underdeveloped areas, for the continuation of our civilisation and its further development, atomic energy is not merely an aid; it is an absolute necessity' (UN, 1955, vol. 16, p. 33).

On being asked if he would opt for nuclear power if it were more expensive than other power sources, Bhabha replied,

'In that case what we would do is to say that we must have a minimum programme and plan, say, for one station at least, so we are up with the technology, prepared for the time when it becomes economical' (*International Science and Technology*, October 1963, p. 93).

Sethna seems to have similar views to Nehru and Bhabha:

'For any development and modernisation programme it is imperative that the people become scientifically oriented before any worthwhile effort at economic development can be undertaken' (*Indian Journal of Power and River Valley Development*, March 1979, p. 49).

Opposition to the nuclear power programme in India has been very limited. Within government circles, the Planning Commission is known to be very sceptical about the claims of the DAE regarding the economics of nuclear power. The planners tend to believe nuclear power is not economic in India at present, but that the future prospects for the technology are good. The early stages of the nuclear power programme were not influenced by the Planning Commission since the DAE was only brought into the five year plan system for the Third Plan (1961/66). Since that date the Planning Commission has tended to accept the DAE's recommendations because (a) the latter body is said to be more technically qualified to pass judgement; (b) there is a desire to avoid internal disputes amongst government offices.

Since at least 1954, the Indian programme has been conceived in terms of a three-stage plan. It is a long term plan aimed at the eventual utilisation of India's large reserves of thorium. The three stages are as follows: (a) operate natural uranium reactors using Indian uranium and producing not only electricity but plutonium; (b) use the plutonium in fast breeder reactors to breed both more plutonium and uranium 233 from thorium; (c) use the uranium 233 to sustain breeders converting thorium to uranium 233.

Although the Indian authorities have been accused of developing their nuclear capability with military intentions, there is in fact something of a conflict between this plan for electricity production and any interest in nuclear weapons. It is true that the CANDU type of reactor is a better producer of weapons-grade plutonium than most other types, and that reprocessing plants are indispensable if it is desired to produce a plutonium bomb. However, it is also true that CANDU reactors and reprocessing plants are key elements in the first stage of this plan. If plutonium were to be used in any quantity for nuclear weapons then the three-stage plan could not grow to any reasonable size. (See the appendix to this chapter for an account of India's nuclear explosion.)

India began its nuclear programme by building several research reactors. Table 3.1 lists the research reactors built in India to date. Only one of these reactors needed substantial outside assistance, and that was CIRUS, which was an almost identical copy of the Canadian NRX reactor. However, it was felt necessary to obtain foreign assistance for the building of power reactors, partly because other countries had more experience, and partly because many of the large scale, sophisticated pieces of equipment were just not obtainable in India.

Homi Bhabha had talks with General K. D. Nichols, Chairman of Westinghouse, in early 1960 about the establishment of nuclear power plants. The USSR offered large scale nuclear assistance to India at attractive terms in March 1960. It is believed that a full size nuclear power project was mooted, but Russian enthusiasm cooled while India displayed interest in commercial tenders from North America and Europe. As a result an agreement concluded in October 1961 between India and Russia referred only to nuclear research.

When it came to power reactors a choice had to be made, firstly between those fuelled with natural or enriched uranium, and secondly between different designs of whichever of these first types was chosen. The obvious choice for India was to go for natural uranium reactors:

(a) This avoided dependence upon a foreign country (the USA essentially) for fuel supplies. It would be possible to produce natural uranium in India but not enriched uranium.

(b) Consequently this avoided long term, foreign exchange commitments which might aggravate India's balance of payments position.

Table 3.1 *Research reactors*

Name	Thermal power	Fuel	Moderator	Coolant	Completion date (criticality)	Notes
Apsara	1 MW	Enriched uranium metal	H_2O	H_2O	August 1956	Swimming pool reactor, built by Indian scientists and engineers, apart from the fuel which was imported from UK.
CIRUS	40 MW	Natural uranium metal	D_2O	H_2O	July 1960	Reactor provided by Canada, heavy water by the USA. India met the costs of construction and provided half the initial fuel load. Full power was reached in October 1963. Used for radioisotope production.
Zerlina	100 W	Natural uranium metal; later, natural uranium dioxide	D_2O	Air	January 1961	Wholly Indian built apart from the heavy water.
Purnima	10 W	Plutonium dioxide	None		May 1972	Zero-energy fast reactor. Indian built.
R-5	100 MW	Natural uranium metal	D_2O	D_2O	1982?	Indian built. To be used for radioisotope production and testing of power reactor fuel.

Sources: DAE Annual Reports; *Nuclear India (passim)*.

(c) Plutonium production, ready for the fast breeder stage
of the programme, would be higher.

There were essentially only two designs of natural
uranium reactor to choose between. These were the Magnox
reactor used by the British and French, and the Canadian
CANDU reactor. At the time that India was about to
embark on a power reactor programme (the early 1960s)
much more experience had been obtained with the Magnox
than with the CANDU. However, India expressed a preference for the CANDU, essentially because it was more
economical on fuel than the Magnox.

For reactors of the same energy output, a Magnox requires about four times as much fuel for its initial inventory
as a CANDU, and correspondingly more uranium for re-
fuelling thereafter. Already it was suspected that India did
not have an abundance of uranium, so the CANDU seemed
a sensible choice.

However, when it came to selecting from the tenders for
India's first nuclear power station, an enriched uranium
boiling water reactor was chosen on the grounds that it was
too good a bargain to miss. It was felt that such an inexpensive turnkey project would demonstrate the economic
viability of nuclear power in India, and thus pave the way for
a full programme. Later, the Estimates Committee of the
Indian Parliament criticised this decision as 'a hasty step,
not in keeping with the country's long-term objective [of
self-reliance]' (Estimates Committee, 1970, p. 30).

In attempting to justify the choice of a BWR, Homi
Sethna has stated that tenders based on the heavy water
reactor system were not received, as at that time none of the
tendering parties had enough experience on this system
(IAEA, 1977, vol. 6). This is untrue, since two out of the
seven proposals were for heavy water-moderated reactors
(see page 41).

Almost at the same time as the decision was being made
in favour of a BWR, negotiations with the Canadians over a
CANDU reactor were making steady progress. So it was
during the same period, 1963–4, that firm agreements were
made on a BWR for the Tarapur site in Maharashtra and
also a CANDU for the Kota site in Rajasthan. It was in-

tended to build more than one CANDU style power station, thus making it the basis of the first stage of the three-stage plan. At that time no commercial size power reactor of this type had been operated even in Canada itself. Canada's Nuclear Power Demonstration Reactor (of 22 MWe net) first went critical in April 1962. It was in August of the same year that the Indian Cabinet decided to approach the Canadians to purchase a CANDU. There was thus no time to learn from NPD's operating experience. Douglas Point (200 MWe) went into commercial operation in Ontario only in September 1968. As has been noted, its performance was generally regarded as disappointing. Nevertheless, the Indian programme is based essentially on this reactor type, RAPS-1 being virtually identical to it.

While the firming up of these proposals for power reactors was in process, design work was going ahead on a prototype power reactor. This was a 15 MWe reactor with natural uranium fuel and moderated and cooled with heavy water, but capable of being operated with other coolants such as light water. It was to provide necessary experience for the design, construction and operation of full sized CANDUs. Its estimated cost was Rs 86.5 million and construction work on it was due to begin in 1965. However, just before construction work was to start the project was dropped, the reason given being that experience gained in the construction and operation of RAPS (with Canadian help) would prove sufficient for developing the rest of the programme. Thus the Indian programme was to begin without anything intermediate between small research reactors and large power reactors. This decision was essentially reversed when, in May 1974, construction work began on the R-5 reactor, the first research reactor in India to resemble the CANDU in using heavy water as moderator and coolant.

Power reactors

Tarapur Atomic Power Station (TAPS)
A government decision was taken in 1958 to build a nuclear power station in the Western Region. The selection of the Tarapur site, 100 kilometres north of Bombay, was

announced in the Indian Parliament in August 1960. (The location of India's nuclear power facilities is shown in Fig. 3.1). This site was chosen because (a) it could supply electricity to the industrial cities of Bombay and Ahmedabad; (b) it was on the coast so that sea water could be used for cooling purposes; (c) it had a low population density. (The population criteria for siting reactors in India are as follows: no population within 2.4 km radius; a low population within 5 km radius; no population centre of 10 000 or more within 16 km and no city of 100 000 or more within 40 km. A village of 600 people within 2.4 km of Tarapur was shifted 5 km away.)

Global tenders were invited for this project in October

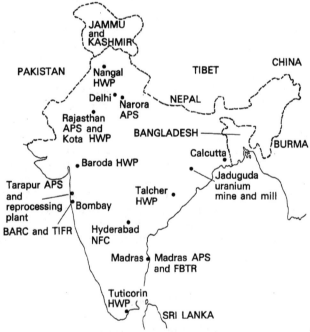

Figure 3.1 India's nuclear power facilities. HWP, heavy water plant; APS, atomic power station; BARC, Bhabha Atomic Research Centre; TIFR, Tata Institute of Fundamental Research; FBTR, fast breeder test reactor; NFC, nuclear fuel complex.

1960. It had originally been intended to instal a single reactor at Tarapur with a capacity of 250 MWe. But it was later decided that this would be too large a unit for the grid to cope with, so that proposals were invited for two reactors of 150 MWe each. Seven tenders were received as shown in Table 3.2. Among the natural uranium reactor proposals the French was said to be the 'best', and among the enriched ones General Electric's was said to be the best. The cost of the French proposal was Rs 890 million (including Rs 593 million in foreign exchange) and the cost of GE's was Rs 607 million (including Rs 442 million in foreign exchange). The tenderers were requested to keep foreign exchange costs to a minimum.

Preliminary agreements to build TAPS were signed in August 1963. It was to be a two-unit BWR of approximately 190 MWe per unit, built by GE, who were to supply the fuel also. Each unit was to be similar to the Dresden-1 reactor which had been built by GE in Illinois and completed in 1960. GE subcontracted Bechtel Corporation to do the con-

Table 3.2   *Tenders for Tarapur Atomic Power Station*

Organisation	Fuel	Moderator	Coolant
GEC (UK)	Natural uranium	Graphite	Gas
English Electric–Babcock Wilcox–Taylor Woodrow (UK)	Natural uranium	Graphite	Gas
Groupement de Constructeurs Française de Centrales Nucléaires	Natural uranium	Graphite	Gas
General Electric (USA)	Enriched uranium	H_2O	H_2O
Westinghouse (USA)	Enriched uranium	H_2O	H_2O
Atomics International/Kuljian Corp. (USA)	Natural uranium	D_2O	Organic
Canadian General Electric and Canadian Bechtel	Natural uranium	D_2O	D_2O

Source: Estimates Committee (1970, p. 27).

ventional engineering. In accordance with conditions laid down by USAID, the Kuljian Corporation of Philadelphia and Nuclear Utility Services of Washington were contracted as consulting engineers to give independent engineering advice on all aspects of the project.

A loan of $80 million was made by USAID toward the reactor cost and the fabrication of the initial fuel charge. This came at an interest of 0.75% and repayment was to be over a thirty year period beginning June 1976. The enriched uranium for the initial fuel charge, worth $14.5 million, was provided on deferred payment terms by the USAEC. The US Export-Import Bank provided the finance to cover this deferred payment period which lasted up to June 1973. The fuel was sold under the same schedule of charges as that for private users of enriched uranium in the USA.

Construction began in October 1964 and GE gave a firm completion date of October 1968, as well as guarantees about the output and efficiency of the station. But a problem arose in 1968 when it was found that cracks had appeared in certain stainless steel components connected with the reactor vessel. This problem also occurred in a similar plant built at Oyster Creek, New Jersey. A programme of repair and modification was carried out at the expense of the contractor. Mainly as a result of this problem the completion of TAPS was delayed to October 1969, in which month it came up to full power and produced commercial electricity for the first time. In compensation for the delay GE paid damages of Rs 1.2 million. However, a bonus of Rs 26 million was paid to GE on account of the net electric output being rated higher than specified in the contract – 400 MWe rather than 380 MWe (Estimates Committee, 1973, p. 11).

Virtually all the equipment for TAPS was imported. The two reactors are housed in a single building, and the equipment layout was kept very compact for reasons of economy, leading to problems in terms of maintenance in later years. TAPS is slightly different to the idealised, conventional BWR. It operates on a dual cycle, that is, a direct cycle of steam to turbine and an indirect cycle using secondary steam generators. About one-quarter of its electrical output is produced by the indirect cycle.

Cooling water for the plant is drawn in from a pipeline

leading directly out to sea and discharged through canals running almost parallel to the coast. Low level radioactive waste is also discharged through these canals. Unfortunately the coastal sea water does not mix very rapidly with offshore water so that the concentration of radioactivity immediately off the coast is rather high. The area around Tarapur is a fishing ground and crabs, oysters, lobsters and prawns are caught in the coastal area.

About 200–300 design changes have been made in TAPS since it began operating. The total capital investment in TAPS by the mid 1970s was Rs 970 million (DAE Performance Budget 1979/80).

To begin with all the fuel for TAPS was enriched and fabricated in the USA. But beginning in 1973 enriched UF_6 supplied by the USA has been converted into UO_2 and fabricated into fuel elements at the Nuclear Fuel Complex, Hyderabad. Thus, since 1975 all the reload fuel for TAPS has been supplied by NFC. Originally, the USA undertook to supply fuel requirements for the whole life of the station but after the testing by India of a nuclear explosive device, in May 1974, supplies of enriched uranium from the USA became irregular. Up to the date of the Indian nuclear explosion, the USA had supplied about 150 tonnes of fuel, including the initial inventory. From May 1974 to May 1980 about 90 tonnes of fuel were supplied, that is rather less than required to run TAPS at full power.

Under the Non-Proliferation Act passed by the US Congress in March 1978, all countries receiving nuclear materials or equipment from the USA are required to accept full scope safeguards on all their nuclear facilities. ('Safeguards' are accounting checks, carried out by the IAEA usually, to detect the diversion of nuclear material to nuclear explosives production. The term 'full scope safeguards' is usually taken to mean safeguards applied to all present and future nuclear facilities in a particular country.) Since India refuses to accept this condition, no more fuel should have been supplied after March 1980. However, by May 1980 it appeared that the US State Department was in favour of selling fuel to India, for this would create closer ties at a time of turmoil in Asia due to Soviet intervention in Afghanistan and the Iranian revolution. The US Nuclear Regulatory

Commission appeared to be against the sale. (The Nuclear Regulatory Commission had earlier warned that if further nuclear devices were exploded, no further fuel would be supplied.) In September 1980 the House of Representatives and the Senate Foreign Relations Committee both voted against the sale of fuel to India. But after intense lobbying President Carter managed to get the Senate to vote in favour, so the export went ahead. However, in July and August 1981 talks were held between the USA and India on the subject of terminating the 1963 Tarapur agreement (and thus ending fuel supply from the USA). As part of an agreed termination the USA wanted some kind of safeguards on the spent fuel (i.e. including agreements not to use it to manufacture explosives or to transfer it to a third country). As an alternative to this it seems the Americans would have accepted the return of the spent fuel to the USA (*Nucleonics Week*, 6 August 1981).

India has never seriously considered building its own uranium enrichment plant because an economic plant would have to be so large (far larger than necessary to supply TAPS) as to absorb an enormous amount of capital to set it up and an enormous amount of electricity to run it. But some thought has been given of late to running TAPS on mixed oxide fuel, that is UO_2 and PuO_2, using the plutonium produced by TAPS and RAPS-1. This would probably only be attempted if fuel supplies to TAPS were irredeemably cut off.

Rajasthan Atomic Power Station (RAPS)
Cabinet approval was given for the building of RAPS-1 (a 200 MWe net reactor) in August 1962. No global tender was issued, instead AECL was invited to submit a proposal for a CANDU reactor, since this was the preferred model for the Indian programme. Agreements on this proposal were signed in April 1964 whereby the Export Credit Insurance Corporation (now the Export Development Corporation) of Canada supplied a loan of $37 million at 6% interest. Eighty per cent of this loan was required to be spent in Canada.

AECL was responsible for the design of the nuclear portion of the plant and Montreal Engineering Company for the design of the rest. Responsibility for commissioning the whole plant lay with AECL essentially, but was subcontracted to Ontario Hydro. Most of the plant was manufactured in Canada. The only major items produced in India were the main transformer and the switchyard. At 250 MVA the transformer was the largest ever produced in India, and was made by BHEL. Apart from these items, India produced some of the smaller components such as instrumentation and piping, and also did some finishing work on the calandria end shields which had been machined in Canada.

Half the initial fuel load was supplied by Canadian Westinghouse and the other half was made in India using fabrication equipment supplied by Canada. The 230 tonnes of heavy water needed as moderator and coolant were to have been supplied by Canada. Due to problems with the Canadian heavy water plants, Canada provided only 130 tonnes from stocks obtained by them from the USA. The USSR supplied a further 80 tonnes in 1973 which made sufficient to commission the reactor.

From the mid 1960s Indian personnel were trained in Canada in the design, construction and operation of this type of plant. Construction of RAPS-1 began in December 1964 and it was expected to go into full power operation in late 1969. In fact it first supplied commercial power to the grid in December 1973 and came to full power in 1974. The capital cost of RAPS-1 was Rs 730 million, which is fairly similar to the capital cost for Douglas Point of $85 million (DAE Performance Budget 1979/80; AECL Annual Report 1966/7).

An agreement was signed to build RAPS-2 (identical to RAPS-1) in December 1966, and construction began in 1967/8. A loan under the same conditions as for RAPS-1 was arranged from the Export Credit Insurance Corporation, this time of $38.5 million. There was much more Indian involvement in the building of this second reactor. The Indian DAE was responsible for its commissioning and a high proportion of the components were made in India. Items which still had to be imported included the turbine and generator, the pressure tubes and some of the calandria tubes, and various valves and pumps. Some of the major pieces which were made in India for the first time, mainly in the private sector, included the calandria, the end shields, the steam generators, the dump tank, and the fuelling machines. However, many important materials required to manufacture these items had to be imported. These included zircaloy, superior grade carbon steels, stainless steel, inconel and monel.

As a result of this increased Indian involvement the foreign exchange cost of RAPS-2 amounted to about 40% of its total cost, in comparison with about 60% for RAPS-1. (These foreign exchange components are uncertain estimates and should be taken as being of the right order of magnitude only.) But some of the indigenously manufactured items took quite a long time to deliver; more than four years for the calandria and five years for the end shields. The delay in delivering these reactor core components set back the schedule for completion of RAPS-2, since the critical path for construction ran through the reactor core. So a major cause of delay in commissioning nuclear power plants in India has been the problem of industrial production. There are signs that this may become less of a problem though, since the number of man-hours needed to fabricate dump tanks and calandria has been decreasing (Table 3.3). However, the time taken to supply the calandria for MAPS-1 was the same as for RAPS-2, and the end shields took just seven months less, so there has not been a marked improvement as yet.

This problem of industrial delay may be contrasted with that in the rich countries advanced in nuclear technology, where regulatory delay has been evident. This has not been a problem in India since the DAE is promoter and regulator of nuclear power developments. Not only does the DAE

Table 3.3   *Man-hours for reactor component fabrication*

Reactor	Dump tank	Calandria
RAPS-2	75 000	120 000
MAPS-1	66 000	85 000
MAPS-2	55 000	65 000

Source: Urja, June 1977, p. 35.

design, build and operate all the nuclear power plants in India, it also has the responsibility for evaluating the performance and safety of these plants.

In 1973 the Canadians halted deliveries of nuclear equipment because of India's refusal to sign the NPT. In 1974, due to the Indian nuclear explosive test, which had used plutonium produced in the CIRUS reactor, Canada cut off all technical assistance and supply of equipment. This test had, according to the Canadians, violated the agreements between the two countries relating to CIRUS, because it had military implications. The Indians argued that it was not contrary to the agreements because it was entirely peaceful in intent. After two years of negotiations the Canadians refused to change their stance and made an official announcement of their complete withdrawal. (Nuclear exports and nuclear assistance to Pakistan were also terminated in 1976.)

The RAPS-2 reactor therefore had to be completed by Indian engineers. It was essentially finished by 1978 but it had not been operated by early 1980 due to a shortage of heavy water. India had received heavy water from the USSR for RAPS-1, and in 1976 it again turned to that source for supplies for RAPS-2. About 55 tonnes (one-quarter of the total needed) were immediately supplied at a price of Rs 1450 per kilogram. However, further supplies were subject to negotiation over the price, and also over safeguards against the heavy water being used to produce nuclear explosives. So, in 1977 the USSR was asking Rs 2000 per kilogram and in November of that year obtained a safeguards agreement through the auspices of the IAEA. This says essentially that the following items shall not be used for the manufacture of *any* nuclear explosive device (peaceful or otherwise): (a) the heavy water supplied by the USSR under the arrangement; (b) RAPS-1 and RAPS-2 (upon introduction of the supplied heavy water); (c) any nuclear material produced from the above two items.

Apparently the USSR pressed India to apply safeguards to their research reactors as well as their power reactors, but the Indians refused. (India's operating power reactors and nuclear fuel complex are subject to safeguards but its research reactors, its planned power reactors, its reprocess-

ing plants, and its heavy water plants are not. Of the
research reactors only CIRUS is capable of producing a
significant amount of plutonium.) Apparently as a consequence of disagreement over price and safeguards, the
USSR has been rather dilatory over further supplies of
heavy water. A consignment of 30 tonnes was despatched in
May 1980, bringing total consignments to around 230
tonnes, sufficient to commission RAPS-2. So, by March
1981 RAPS-2 was being test run at 130 MWe.

The capital cost of RAPS-2 by March 1979 was about
Rs 920 million.

Reactors under construction and planned
Construction of two further CANDU reactors at Kalpakkam near Madras (MAPS) began in 1969–70, and two
more at Narora (NAPS) in Uttar Pradesh in 1975–7. None
of these were as yet in operation by January 1982 but
MAPS-1 was likely to be operable in 1982, if heavy water
was available. There is now no foreign collaboration in the
construction of any of these reactors. Some materials, e.g.
special steels and nickel alloys, had to be imported, but all
the major items of equipment are being made in India,
including the turbines and generators, which are to be
supplied by BHEL. Pumps are one of the few items which
still have to be imported. The proportion of foreign exchange in the capital costs of these plants is expected to be
around 15–20% for MAPS and 9–15% for NAPS.

All of these reactors are slightly modified versions of the
RAPS design. They each should have a slightly higher rated
output than RAPS (235 MWe gross rather than 220).
Whereas RAPS has a dousing system to limit pressure rise
after a loss of coolant accident, MAPS and NAPS are to
have suppression pools. MAPS and NAPS are being fitted
with dump-condensers, while RAPS was not. (Excess steam
from the reactor is diverted to the dump-condenser when-
ever there is a sudden reduction in power demand, thus
avoiding sudden alteration of the reactor core conditions to
match the demand. This should reduce fuel failures.) NAPS
is to have cooling towers, whereas TAPS and MAPS use
sea water, and RAPS lake water, for cooling. Also, NAPS is
to have four boilers and four primary heat transport pumps

rather than eight, in order to scale up these components to a size suitable for a 500 MWe reactor. The latest estimate of the total capital investment in the two plants is MAPS 1 and 2, Rs 2109 million, and NAPS 1 and 2, Rs 3115 million. Plans have been announced for the building of five more nuclear power stations each to consist of two reactors of 235 MWe. The first of these is to be sited at Kakrapar, in the state of Gujarat, which is on the river Tapti, 100 km from Surat. The projected total cost has officially been put at between Rs 3600 million and Rs 3700 million. After this stage of the programme is complete (i.e. after ten more Narora-style reactors), it is planned to introduce a 500 MWe heavy water reactor on which a design group was working by 1981 (*Nucleonics Week*, 16 July and 5 November 1981).

Uranium mining and exploration

India began looking for uranium in 1948 when the AEC was established. By 1954 it was realised that uranium ore in India was likely to be of low grade, with about 0.06% of uranium being the best then discovered, and most ores being considerably worse than this. By 1964, when it was definitely decided to embark on a power programme, resources of uranium were thought to be enough to fuel 3000 MWe for 25 years. Despite spending Rs 458 million on uranium exploration over the past thirty years, the resources of uranium in India remain fairly sparse. (Of the non-Communist world only Australia, Brazil, Canada, France and the USA have spent more on uranium exploration in this period.) Reasonably assured resources amount to about 29 000 tonnes of uranium while estimated additional resources come to about 24 000 tonnes. The assured resources are sufficient to support a programme of about 5000–6000 MWe of CANDUs for a period of thirty years. (Total reserves of uranium (assured and estimated) are equivalent to less than 1% of India's estimated coal reserves.) All of this uranium is in low grade ore, 0.06% or less, whereas exploration world wide has been directed towards ore of grade 0.1% or greater. As a result India's uranium is about twice as expensive as uranium from Canada.

The first source of uranium exploited in India was the beach sands of Kerala. Uranium from this source provided some of the initial fuel for CIRUS in 1959/60. These beach sands contain from 1.5 to 3.5% of monazite, which is a source of rare earths. The monazite in turn contains 0.2–0.6% of U_3O_8. This is, overall, a very low grade of uranium content, thus the production of uranium can only be carried out as an adjunct to monazite/rare earth production.

Uranium deposits at Jaduguda in Bihar State were discovered in 1950. The exploitable deposits there amount to about 3600 tonnes of uranium. A mine and a mill were opened at Jaduguda in 1968. The mill is capable of producing 200 tonnes of U_3O_8 per year. In the mill, the ore is crushed and ground, and uranium is leached into solution by adding sulphuric acid. Uranium from the solution is precipitated in the form of magnesium diuranate and this concentrate of uranium is sent to the Nuclear Fuel Complex for further processing. The combined capital cost of Jaduguda mine and mill was about Rs 107 million.

Shortage of ore in the upper levels of Jaduguda mine has led to the need for deeper shaft sinking. It is projected that a new uranium mine and mill will be established to come into operation by about 1985/6 at a cost of Rs 243 million.

On at least two occasions during the 1970s, smuggling rings dealing in uranium concentrate from Jaduguda have been broken by the police. The uranium was supplied via Nepal or Hong Kong to Chinese and Pakistani agents. The amounts of uranium recovered were very small, but it has been suggested that the amount stolen was of the order of tens of tonnes (Lovins, 1975, p. 35; *Nuclear Engineering International*, January 1979, p. 10).

Fuel fabrication

A small plant for the production of uranium metal fuel elements went into operation at BARC in 1959. This supplied fuel for CIRUS. A decision was made in the late

1960s to set up a Nuclear Fuel Complex to convert enriched UF_6 and fabricate it into fuel for TAPS, and also to fabricate natural uranium oxide fuel for RAPS. The estimated capital cost for this was put at Rs 138 million in 1970. The complex was built near Hyderabad and came into operation during 1973–75. The enriched fuel fabrication plant is designed to meet the reload requirements of TAPS and is capable of producing 24 tonnes of fuel per year. The natural uranium fuel fabrication plant is capable of producing 100 tonnes of fuel per year and this is reckoned sufficient to provide the reload requirements of three or four 200 MWe CANDUs. NFC is also able to produce, entirely from indigenous resources, 50 tonnes of zircaloy per year for fuel cladding – sufficient to set up two 200 MWe CANDUs per year. The total capital outlay on NFC was estimated in 1979 to have been Rs 1074 million. The foreign exchange component of this outlay was less than 30%, mainly for the importation of materials such as inconel and stainless steel and for a Soviet-built extrusion press.

Heavy water plants

India's first heavy water plant at Nangal in the Punjab was built by a West German contractor, Linde GmbH, and was commissioned in 1962. It is associated with a fertiliser plant and uses a hydrogen distillation technique to produce about 14 tonnes of D_2O per year. This technique is fairly power intensive since refrigeration is required to maintain the necessary low temperature. The cost of heavy water from this plant was, in 1974, around Rs 1000 per kilogram, whereas the import price was around Rs 800. The plant is large enough only to provide for research reactors or small top-up requirements for power reactors. India has therefore embarked on the building of four large scale plants whose details are shown in Table 3.4.

All of these plants require several tall and heavy exchange towers. The largest is at Baroda, weighing 530 tonnes and measuring 33 metres in height. All of the exchange towers have been imported except those for the Kota plant. Around 70–90% of the components for this latter plant have been

Table 3.4 *Heavy water plants*

Location	Contractor	Process	Capacity (tonnes per year)	Start of construction	Current estimated cost (Rs million)	Notes
Baroda	GELPRA (one Swiss + two French firms)	NH_3/H_2 exchange, monothermal	67	1969	401	Very high pressure process (about 650 atmospheres)
Kota	DAE, India	H_2S/H_2O exchange, bithermal	100	1970	559	20 atmospheres
Tuticorin	GELPRA	NH_3/H_2 exchange, monothermal	71	1971	374	Similar to Baroda
Talcher	Friedrich Uhde GmbH (West Germany)	NH_3/H_2 exchange, bithermal	65	1972	508	Very high pressure process (about 300 atmospheres)

Sources: DAE Annual Reports; *Nuclear India (passim)*; *Commerce*, 6 December 1975, Supplement, pp. 1–20.

indigenously manufactured, while 30–40% of the components for the other plants have been made in India. All of the ammonia–hydrogen exchange plants have been set up as adjuncts to fertiliser plants.

Most of the world's heavy water has been produced in the USA and Canada using the Girdler sulphide H_2S/H_2O exchange method. However the USA and Canada were not willing to share the know-how on this technology with India, therefore the Indian DAE had to redevelop the process themselves to design the plant at Kota. Its construction was taken on without first building a pilot plant from which to gain experience on a limited scale (Mahatme, 1975, p. 691).

Only one plant using the ammonia–hydrogen exchange method has been operated in the world. This was a 20 tonne per year plant built by GELPRA at Mazingarbe in France, which operated from 1968 to 1972.

All of the Indian heavy water plants have been plagued with difficulties. Despite being expected to function within four to five years of the start of construction, none of them have as yet produced a significant quantity of heavy water. The Baroda plant suffered an explosion while undergoing trial runs in December 1977 which put it out of action until 1981. The cause of the accident has been laid down to a design deficiency. The Kota plant had not yet operated by late 1981. The Tuticorin plant was operated for only a few weeks during 1978 and 1979 due to technical problems and a labour dispute. The Talcher plant has been delayed by two years due to the loss of two exchange towers, which reportedly fell into the sea off Portugal from the ship which was transporting them in 1975 from Germany.

The DAE has announced that it intends to set up three more heavy water plants. Two of them are to be run in conjunction with fertiliser plants being set up by the Fertiliser Corporation of India, while the third is to be similar to the one at Kota. It is hoped that the first two heavy water plants will be completed by 1986/7, and the third by 1989/90.

Reprocessing plants

India's first reprocessing plant was built at BARC between 1961 and 1964. It was entirely Indian built and cost about

Rs 35 million. It was capable of reprocessing 30 tonnes of uranium metal fuel per year using the Purex process, and of extracting the plutonium. The plutonium for the Indian nuclear explosion came from CIRUS fuel which had been reprocessed at this plant. The plant was shut down soon after the 1974 nuclear explosive test, to be decontaminated and expanded so as to be capable of reprocessing fuel from the R-5 research reactor. The plant is said to have been decontaminated, and should be ready to resume expanded operations in about 1982.

Construction of a 'power reactor fuel reprocessing plant' at Tarapur was begun in 1969. It is designed to reprocess oxide fuels from TAPS and RAPS and has a capacity of 100 tonnes of uranium per year. It too uses the Purex process. The plant became physically operational in 1977 and cost about Rs 100 million to build. A first trial run with spent fuel was completed in 1979, and in April 1981 plans were afoot to reprocess RAPS-1 spent fuel on a regular basis.

By early 1981 the plant had not been used to reprocess spent fuel from TAPS due to American refusal to permit this. Under the terms of agreement over TAPS, India was to be allowed to reprocess fuel subject to 'joint determination' of the two governments. And, furthermore, the plutonium obtained could be used by India as fissile material in the reactors, if the USA did not take up its option to purchase the plutonium for its own peaceful purposes. It seems that, despite problems with domestic reprocessing plants and waste storage facilities, in 1976 the USA demanded the return of spent fuel from TAPS if nuclear fuel supplies were to continue. India is said to have reluctantly agreed to do this in 1977, but by early 1980 no spent fuel had actually been returned. One reason why the USA might have wished to back-pedal on this issue was the fact that under the 1963 Tarapur agreement any fissile material returned to the USA was guaranteed *not* to be used for weapons purposes. Thus, if spent fuel were returned to the USA, India should have the right to inspect the facilities which handled the fuel. A situation in which a non-nuclear weapons state makes inspections on a nuclear weapons state could be anticipated to be diplomatically awkward (and possibly also very revealing).

The spent fuel storage pond at TAPS had been filled to beyond its design capacity by 1977. Extra storage racks have been placed in the pond sufficient to absorb spent fuel up to 1980/1. In all about 160 tonnes of spent fuel was stored at TAPS by late 1979. Theoretically it should be possible to use this spent fuel as a fuel supply for CANDU reactors, since the uranium 235 content is still up around the natural level (0.7%). In this way it would be possible to obtain additional energy without recourse to reprocessing. Whether India has investigated the feasibility and economic worth of this approach is not known. There would, of course, be some problems in handling such radioactive fuel.

The highly active wastes generated by the Tarapur reprocessing plant are to be immobilised in a solid vitreous (borosilicate) matrix. Work began on this waste immobilis-ation project in 1973 and construction work was in progress in 1976. It was hoped to commission the facility in 1980. After solidification it is planned that the waste will be stored temporarily in air-cooled underground vaults (preliminary designs for these were completed in 1978). After twenty or thirty years of this interim storage the decay heat should have reduced sufficiently to allow permanent geological disposal. Evaluation of sites for permanent disposal was under way in the late 1970s. A dumping site somewhere in Andhra Pradesh has apparently been proposed.

Other aspects of the programme

The DAE has involved itself in the research, development and use of radioisotopes. Topics covered include radi-ography, process control, tracers for leak detection, medical sterilisation, static electricity dissipation, smoke alarms, food preservation, insect sterility, and mutation breeding of agricultural crops. Many of these techniques are in use in India today. Most of the radioisotopes are produced at the CIRUS reactor and their sales have amounted to between Rs 7 million and Rs 8 million per year recently. (This is an insignificant amount in comparison with electricity sales from TAPS, say.)

BARC is India's largest scientific establishment, employing well over 10 000 staff. All the research reactors are sited there and the research programme includes such diverse topics as magnetohydrodynamic power generation, photochemical reactions, bacterial genetics, immunology and desalination, as well as subjects more strictly related to the nuclear. power programme. The level of sophistication of work carried out there is high. Support services for the research programme are excellent.

Appendix: India's nuclear explosion of 1974

India has said that the underground nuclear explosion of 18 May 1974 was for peaceful purposes only, designed to discover the effects on the surrounding earth and rock with a view to using such explosions for excavating harbours and canals, and for mining low grade minerals. The concept of using nuclear explosions for peaceful purposes seems first to have arisen in the USA, where the Plowshare Project to develop such uses was initiated in 1957. This project was terminated in 1970, by which time $138 million had been spent (Commoner, 1972, p. 60). The USA and the USSR seem to be the only countries to have taken a significant interest in the topic. By 1971 the USA had conducted fourteen peaceful nuclear explosions (PNEs) and the USSR seventeen. The USA last detonated a device to investigate the feasibility of PNEs in 1973, and in 1977 it was announced that no further work would be done on this topic. In the same year, while discussing a Comprehensive Test Ban Treaty, the USSR offered permanently to suspend experiments with PNEs. The USSR persevered with PNEs when no treaty was signed, however, exploding twenty-two devices between 1975 and 1979.

The Indian government has, over the years, fairly consistently stated that it is not interested in acquiring nuclear weapons. But for many years there has been an articulate lobby in favour of the bomb. A number of 'incidents' have provided fuel for this lobby's arguments, including the border dispute with China in 1962, China's first nuclear explosion in 1964, and military conflict with Pakistan

Table 3.5   *Prime Ministers' attitudes to Indian nuclear explosions*

Prime Minister	Date of office	Attitude
J. Nehru	1947–64	Never
L. B. Shastri	1964–66	Not at present
I. Ghandi	1966–77	Keep the option open
M. Desai	1977–79	Against

Sources: Strategic Analysis, November 1977, pp. 1–2; *Urja*, 19 August 1978, p. 5.

(particularly in 1965 and 1971). The attitude of successive Prime Ministers to the development of nuclear explosions by India has, in simple terms, been as shown in Table 3.5.

On 17 May 1970 the Chairman of the AEC, Vikram Sarabhai, announced that India would not develop nuclear weapons but would keep the option of underground PNEs. In the same year Canada sought an assurance from India that plutonium from the CIRUS reactor would not be used in explosive devices of any kind. However, as noted in Chapter 3, this *was* the source of the explosive device detonated in 1974.

The experiment took place in the Rajasthan desert at a depth of 107 metres using between 10 and 15 kilograms of plutonium. The yield achieved was about 12 kilotonnes, producing a crater 100 metres across and 10 metres deep. The cost of the explosion was put at Rs 3.2 million. On 19 May 1974 Jagjivan Ram, Minister of Defence, stated that the decision to go ahead with a PNE was made in 1971. Relations with the USA and with Pakistan were particularly poor in that year. World reactions to the explosion were varied. Japan, Pakistan, Canada, the USA and Britain were all critical, while some Third World countries and the USSR felt that it struck a blow for the developing world. (The USSR has since become less favourable to such events.) China remained non-committal on the issue.

India has not exploded any further devices since the first, and there has been no official indication of any continuing work on this area. (In fact between 1974 and 1979 no reprocessing plant was being operated in India and so no further plutonium was available beyond that produced by the BARC reprocessing plant during its ten year period of operation.) But the fact that India also has a programme for the development of rockets for satellite launching makes many observers suspect that India has had strategic military considerations in mind all along. It would, presumably, not take very long to link the parallel space and atomic energy programmes so as to produce a credible nuclear weapon if necessary.

India's decision to explode a nuclear device must be seen within the context of that country's attitude to the Non-Proliferation Treaty (NPT). This treaty, which came into force in 1970, was sponsored by the USA, the USSR and the UK. These countries, in exchange for a commitment by non-nuclear weapons states to refrain from acquiring such weapons, agreed to the following principles (amongst others): (a) not to assist any non-nuclear weapons state to acquire such weapons; (b) to seek the discontinuance of all nuclear weapons tests, including underground tests; (c) actively to assist all signatories to acquire the equipment and information necessary to exploit the peaceful uses of nuclear energy; (d) to make available to all signatories the potential benefits from PNEs; (e) to end the nuclear arms race at the earliest possible date and to move toward complete nuclear disarmament.

Many countries have signed and ratified the treaty, but India has not, along with China, France, Argentina, Bangladesh, Brazil, Chile, Cuba, Israel, Pakistan, South Africa and Spain. Some of the major criticisms of the NPT that have been expressed by these countries are as follows:

(a) The non-nuclear weapons states have specific obligations under the treaty, such as foregoing nuclear weapons and accepting safeguards inspections, while the nuclear weapons states only give vague commitments. Thus vertical proliferation (the acquisition of more and more weapons by the nuclear weapons states) continues while horizontal proliferation (the acquisition of nuclear weapons by non-nuclear weapons states) is banned.
(b) The right to develop PNEs is reserved to the nuclear weapons states.

Therefore an added reason for exploding a nuclear device in India was as a snub to the NPT.

Thus probably the most realistic interpretation of the reasons behind India's nuclear explosion pertains to its political implications rather than its strictly military uses. Granted that the 1974 explosion did not provide much new scientific or engineering information, it can be said that it achieved the following purposes:

(a) It bolstered the Indian government's domestic prestige.
(b) It appeared to give India a greater degree of equality with China.
(c) It provided a perceived superiority over Pakistan.

CHAPTER 4

Assessment of the Nuclear Power Programme

Rate of growth of capacity

Expectations of the growth of nuclear power capacity in India have not been fulfilled. Table 4.1 compares some official forecasts with the actual capacity.

The original forecasts of 1954 predicted a long gestation time (and on this has been accurate enough) but expected the programme to grow rapidly in the 1970s. In fact, during the course of the Fifth Plan, operational nuclear capacity has not grown at all. These Indian forecasts may be compared with the forecast of the IAEA *Market survey for nuclear power in developing countries* prepared in 1974. This anticipated an installed capacity of 4200 MWe of nuclear power in India by 1980. It was, obviously, over-optimistic.

By 1979 the total electrical generating capacity in India was about 28 000 MWe, thus about 2% of the total consisted of nuclear capacity. Clearly then, nuclear power is not yet of major practical significance in India.

Financial expenditures

The total annual budgets of the DAE are indicated in Table 4.2. These figures cover Plan as well as non-Plan expenditures, as do all subsequent figures relating to the DAE. Some of the expenditure figures tabulated in this chapter are approximate (particularly those for recent years, which are only estimated) but they are certainly the most reliable

Table 4.1 *Growth of nuclear power capacity (in electrical megawatts): expectations and actual*

	AEC (1954)	Bhabha and Dayal (1962)	Energy Survey Committee (1965)	AEC (1968)	Fifth Plan (1973)	Sixth Plan (1978)	Actual
1970/1	600		600	400			400
1973/4							600
1975/6	3000		2000	1000			600
1978/9					1285		600
1980/1	8000		5000	2700			600
1982/3						1565	
1987		20 000–25 000					1000?

Sources: as listed, plus Chitale and Roy (1975, p. 97).

Nuclear Power in India

Table 4.2   *Indian DAE total budget*

Year	Rs million
1963/4	129
1964/5	259
1965/6	432
1966/7	
1967/8	587
1968/9	
1969/70	
1970/1	
1971/2	1050
1972/3	1260
1973/4	1180
1974/5	1640
1975/6	1935
1976/7	2155
1977/8	2077
1978/9	2633
1979/80	2950

Sources: DAE Performance Budgets; Rao
(1974); Kaul (1974); *Nuclear Engineering
International*, April 1979, p. 9.

figures to be made public. Any corrections are likely to
involve upward adjustments.

Table 4.3 shows how these budgets, aggregated into Plan
periods, compare with capital investment in other sources of
power over the same periods. It can be seen that the DAE
budget is quite significant in comparison with the capital
expenditure on thermal power and hydropower.

Table 4.3   *Comparison of energy investments*

	Capital investment in all conventional power projects (Rs million)	Capital investment in coal industry (Rs million)	DAE total budget (approx.) (Rs million)
Third Plan (1961/6)	13 340		1000
Three Annual Plans (1966/7, 1967/8, 1968/9)	11 940		1800
Fourth Plan (1969/74)	29 320		5000
Fifth Plan (1974/9)	72 940	11 266	9930

Sources: Wagle and Rao (1978); Swayambu (1972); Sethna and Srinivasan
(1977); DAE Performance Budgets; *Urja*, 10 June 1978.

It is interesting to compare the DAE total budget with that of the UKAEA at a similar (i.e. early) stage in the history of the UK nuclear power programme. To avoid the problem of differing purchasing powers for the rupee and the pound sterling, it is convenient to make the comparison through calculating nuclear expenditure as a percentage of national income. Thus, in 1954/5, two years before the first nuclear power station went into operation in the UK, the UKAEA was spending 0.38% of the UK GNP. In 1967/8, two years before TAPS went into operation, the Indian DAE were spending 0.19% of the Indian GNP. Similarly, in 1958/9, two years after Calder Hall went into operation, the UKAEA was spending 0.56% of the UK GNP. While in India in 1971/2 0.24% of the GNP was spent on the nuclear programme. Therefore, the UKAEA was spending *relatively* twice as much in terms of proportion of GNP as was India. There are at least two possible explanations for this difference – and they are not necessarily mutually exclusive. First,

Table 4.4   *DAE research and development expenditure*

Year	Rs million
1963/4	111.5
1964/5	143.4
1965/6	200.5
1966/7	
1967/8	
1968/9	253.7
1969/70	267.8
1970/1	287.2
1971/2	206.1
1972/3	197.3
1973/4	244.6
1974/5	418.8
1975/6	539.4
1976/7	608.8
1977/8	611.2
1978/9	609.9

Sources: DAE Performance Budgets; *Nuclear India,* December 1978; Department of Science and Technology (1977, 1978); UNESCO (1972, p. 114); Rahman (1974, p. 53).

the UKAEA was involved in a substantial amount of expensive weapons-related expenditure while the Indian DAE was not. Second, the R & D costs for the UKAEA, as a pioneer in the field, were much higher than they were for India, coming in twelve years later.

Expenditures of the DAE on R & D are shown in Table 4.4. It can be seen that R & D absorbed about half of the total DAE budget during the early 1960s, but these days it absorbs one-quarter. As a proportion of total central government spending on R & D, the expenditure on nuclear R & D is very high. Table 4.5 shows that the DAE absorbed around one-quarter of the Indian government's R & D funds during the late 1950s and the 1960s, and nearly one-fifth during the 1970s. Since government R & D accounts for

Table 4.5 *Proportion of Indian Government R & D going to atomic energy*

Year	Total central government R & D (Rs million)	Percentage going to DAE R & D
1955/6	121	14.3
1956/7		
1957/8		
1958/9	277	28.1
1959/60		
1960/1	299	25.7
1961/2		
1962/3		
1963/4	487	22.9
1964/5	586	24.5
1965/6	791	25.3
1966/7	881	
1967/8	930	
1968/9	1110	22.9
1969/70	1232	21.7
1970/1	1449	19.8
1971/2		
1972/3		13.2
1973/4		15.1
1974/5		19.5
1975/6	2500	21.6
1976/7	3200	19.0

Sources: UNESCO (1972, p. 52); Rahman (1974, p. 54); Nanda (1977, p. 66); Department of Science and Technology (1977); Agarwal (1979).

most of the R & D spending in India (i.e. around three-quarters or more), we can see that the nuclear programme is absorbing a very large part of total R & D spending in India. When we realise that, along with most other underdeveloped countries, India spends a very small part of her GNP on R & D (less than 0.5% as compared with, say, 2–3% in western Europe and the USA), it seems clear that India is placing a lot of her eggs in one basket. For, what little amount she does spend on R & D is substantially devoted to nuclear research. In fact, the overall order of priority given to nuclear R & D seems to be roughly similar these days in the UK and India. In 1977, the UKAEA's expenditure on R & D amounted to about 0.09% of GNP, while in the same year the Indian DAE's R & D spending came to 0.08% of GNP.

The Indian Government spends far less on R & D for the conventional power industry than it does on nuclear power. The combined R & D expenditure of the government's Department of Coal, Bharat Coking Coal Ltd and the Neyveli Lignite Corporation Ltd in 1974/5 amounted to Rs 6 million, and in 1976/7 to Rs 19 million. In the same years the combined R & D expenditure of the Department of Power and Bharat Heavy Electricals Ltd (another government-owned firm) came to Rs 16 million and Rs 52 million respectively. Thus R & D for nuclear power absorbs about ten times as much as R & D for the conventional power industry, including the coal industry as a whole.

Reactor performance

Capacity factors
A key criterion for performance of a power station is its capacity factor, which is a measure of actual power generation against the maximum possible generation for the year. Prior expectations of nuclear capacity factors were of the order of 80% (Bhabha, 1957, p. 184; Bhabha and Dayal, 1962, p. 4159). In fact it has been stated (by a Chairman of the AEC) that nuclear power would not be economic in India if a minimum of 75% were not achieved (Sarabhai, 1969, p. 36).

Table 4.6   *Capacity factors for TAPS*

Year	Capacity factor (%)
1969 (November and December)	53.4
1970	59.2
1971	48.7
1972	23.6
1973	54.5
1974	37.0
1975	52.9
1976	62.2
1977	61.1
1978	56.8

Source: IAEA (*Operating experience* . . . , 1969–80).

The combined capacity factors for the two units at TAPS have been as shown in Table 4.6

Since the second half of 1977 both reactors at TAPS have been restricted to about two-thirds of their rated output. This measure has been taken due to uncertainty over fuel supplies from the USA and the problem of the filling to maximum capacity of the spent fuel storage pool. Capacity factors since 1977 have been based on the restricted output as a maximum output.

Getting spare parts imported is a major problem for various reasons, including obsolescence. Repairing rather than replacement has had to be done to a much greater extent than normal.

The performance of TAPS has been affected by the size of grid into which it feeds. The lack of spinning reserve capacity in this grid has frequently compromised satisfactory maintenance. Operation at off-standard frequency is not unusual due to the inadequate capacity of the grid; and fluctuations in demand for TAPS' power has contributed to fuel failures.

Capacity factors for RAPS-1 are shown in Table 4.7.

During the period from its completion up to 1981, RAPS-1 has not functioned in one stretch for more than three months. The reasons for these low capacity factors were as follows:

(a)   Problems with the turbine designed by English Electric and manufactured by Canadian General Electric. (This

Table 4.7   *Capacity factors for RAPS-1*

Year	Capacity factor (%)
1973	17.2
1974	36.8
1975	33.2
1976	44.3
1977	26.4
1978	9.2

Source: IAEA (*Operating experience* . . . , 1973–80).

type of turbine had not been made in Canada before.) In 1973 the bearings were causing trouble, then there were three failures of the turbine blades in 1974–6. These four outages amounted to a total downtime of 14 months. The problem of blade failures was solved by removing some of the high pressure blading, thus reducing output from 200 MWe net to about 175 MWe net. (Douglas Point has also suffered from turbine blade failures, this being one reason for converting to steam supply rather than electricity supply.)
(b) A strike in 1977–8 kept RAPS-1 out of action for a year.
(c) The small size of Rajasthan grid into which RAPS-1 fed. When RAPS-1 was started up, the Rajasthan grid had a non-nuclear capacity of 334 MWe and the next sized generator was one-fifth the size of RAPS-1. Under the circumstances it was to be expected that grid collapses would occur, especially if any problems (however transitory) occurred with RAPS-1. Now a problem unique to nuclear reactors is that if they are suddenly taken off load when they have been operating at a normal power level, they must be restarted and brought up to two-thirds or more of their previous operating level within about thirty minutes, or they cannot be restarted for about forty hours. The reason for this is that neutron-absorbing decay products build up in the core unless the neutron flux is quickly restored. This phenomenon is called 'poisoning out'. This occurred several times at

RAPS-1, because the power needed to start up the reactor (14 MWe) could not be made available again within half an hour. A big step was made toward solving this problem in 1975 when RAPS-1 was connected into the Northern Region Electricity Board system with a capacity of over 5000 MWe. However, this grid system is still not as strong as it might be. Off-standard frequencies and voltage fluctuations continue to affect operation of RAPS-1, and there have still been some grid-induced outages.

The small size of grid into which the Indian reactors feed has been a considerable factor in their performance. According to H. N. Sethna,

'In our country power systems are in a primitive state. The regional power grids are yet to become a reality and the national network a future possibility. We have experience of what damage such power systems can cause on sophisticated nuclear plant. We have had several instances where there has been a total loss of power at our nuclear stations. The plants survived because of the emergency cooling system powered by diesel engines' (DAE, 1976, p. 5).

(The emergency core cooling system at RAPS-1 was triggered in December 1980. Four hundred thousand gallons of water were released in under a minute and the resultant outage lasted a month (*Nucleonics Week*, 26 February 1981).)
The overall capacity factor of TAPS-1 and TAPS-2 has been 50% from 1969 to 1980, and of RAPS-1 40.6% from 1973 to 1980. This gives a combined capacity factor for Indian nuclear reactors of 46%.
From 1969 to 1978 hydroelectric power stations had a capacity factor of 44.5% and thermal power stations a capacity factor of 44.7% (Wagle and Rao, 1978; *Monthly Abstract of Statistics*, July 1977, November 1978). It is a policy to use hydropower for peaking purposes, and some of the thermal stations are gas- or oil-fired and not intended for base-load. Since nuclear power *is* intended for base-load, it is clear that the performance of the nuclear reactors is rather poor in

relation to previous expectations *and* the performance of other power stations in India.

How does the performance of India's nuclear reactors with regard to capacity factors compare with that of their equivalents in the USA and Canada? Over the period 1960–77, Dresden-1 has had an average capacity factor of 50%, which is strikingly similar to that of TAPS. We may deduce therefore that the level of competence in India at running this kind of station is not entirely dissimilar to that in the USA.

It is more difficult to compare the performance of Douglas Point with that of RAPS since the former only operated as a producer of electricity, primarily, for abour four years. During this period (1968–71) Douglas Point had an average capacity factor of 41%. In 1972 there was a shortage of heavy water in Canada and so Douglas Point's heavy water was removed for use in the Pickering nuclear reactors which were then starting up. As a result Douglas Point's capacity factor for 1972 was very low, 18.5%.

In 1973 Douglas Point was converted primarily to supplying steam for Ontario Hydro's Bruce heavy water plant, plus a much reduced electricity output. Between 55 and 70% of the station's energy output went as steam to Bruce. Bruce uses the Girdler sulphide method of heavy water production and this requires an operating temperature of about 130 °C. Thus the steam supply from Douglas Point can be of substantially lower temperature than the normal temperature used for electricity production (about 290 °C). However, it is the practice of AECL to calculate capacity factors for Douglas Point by including the delivered steam output. On this basis the average capacity factor for 1973–77 has been 59%. Overall, the output from Douglas Point has probably been a little better than that from RAPS-1. This difference is possibly explained by the reasons given for the low capacity factors at RAPS-1 stated on pages 66–7. Even if RAPS-1 had performed at a similar level to Douglas Point, it would not look particularly favourable in comparison with the performance of coal and hydro stations or in comparison with expectations.

During Vikram Sarabhai's chairmanship of the AEC. studies were made of the possibility of improving the performance of nuclear power by establishing nuclear-powered agro-industrial complexes. These would have involved large scale production of such items as nitrogenous and phosphatic fertilizers, aluminium, and desalinated water. Through concentrating power-intensive industries around a large 1200 MWe power station, it seemed feasible to raise capacity factors and avoid the problems associated with small grids. Possible sites were proposed in Gujarat and Uttar Pradesh and the capital cost of these complexes was estimated in 1970 at Rs 6000 million and Rs 10 000 million respectively. The large size of power station involved, and the extremely large capital investment necessary, seem to have led to a waning of interest in this approach. Its economics remain somewhat doubtful (Thomas *et al.*, 1970).

Radiation exposures

A feature of TAPS performance has been the high levels of occupational exposure to radiation. Permanent workers at TAPS currently receive, on average, external doses of about 3 rems per year. Originally TAPS was manned by about 250 operating and maintenance staff, but now 1000 people are employed there and nearly 800 of these are radiation workers. The reason for this expansion in workforce has been the need to keep occupational radiation exposures below the ICRP recommendation of 5 rems per year. (The limit set by the International Commission on Radiological Protection for members of the public is 0.5 rem per year and for adults occupationally exposed to radiation 5 rems per year. These limits may be compared with typical natural background radiation in the UK of around 0.13 rem per year, and with an average dose from medical uses of radiation of between 0.02 and 0.07 rem per year.) But even this expanded workforce is insufficient to cope with radiation work. Over recent years 1500 to nearly 3000 temporary workers have been brought in *each* year to deal with maintenance problems. These workers have received, on average, just under 1 rem each.

Table 4.8 shows occupational radiation exposures over recent years at TAPS. It will be noticed that the total exposure has been rising steadily. Table 4.9 shows some figures relating to gross over-exposures to radiation.

Table 4.8   *External radiation exposures to workers at TAPS*

Year	No. of permanent workers	Dose to permanent workers (man-rem)	No. of temporary workers	Dose to temporary workers (man-rem)	Total nos	Total dose (man-rem)
1969					399	43
1970					550	153
1971					622	444
1972					1503	2455
1973					1883	2733
1974	609	1847	1624	1431	2233	3278
1975	689	2248	2198	1996	2887	4244
1976	675	2181	2465	2320	3140	4501
1977	789	2252	2811	2684	3600	4936
1978		2313		2561		4874

Sources: BARC (*Summary of personnel radiation exposures* . . , 1974, 1976, 1977); IAEA (*Operating experience* . . , 1980).

It is normal practice worldwide to keep exposures far below the ICRP limit of 5 rems per year. These exposures at TAPS are very high in comparison with such conventional practice. There are two reasons for added concern about these Indian exposure levels. First, they relate only to exposure to external radiation and not radiation from ingested or inhaled particles. The Indian DAE has estimated that 10% of annual radiation exposure would be due to internal exposure. The external exposures tabulated should therefore be raised accordingly to indicate a true exposure rating. Second, the ICRP recommended limit is based on a 'standard man' weighing 70 kg, drinking 2 litres of water per day, and breathing about 18 cubic metres of air per day. The average Indian is much more vulnerable to radiation expos-

Table 4.9   *External over-exposures at TAPS* (annual)

Year	No. of workers exposed to more than 5 rems	Maximum individual dose (rems)
1974	51	6.36
1975	190	6.27
1976	32	18.24
1977	16	16.19

Source: BARC (*Summary of personnel radiation exposures* . . . , 1974, 1976, 1977).

ure, since he weighs around 50 kg, has a water intake of 4 litres per day and breathes about 25 cubic metres of air per day (*Nuclear India*, June 1970, p. 8).

For all light water reactors in the USA the average radiation exposure levels per unit of electricity produced have been of the order of 1.1–1.8 man-rems per MWe-year. For TAPS from 1974 to 1977 exposure levels have risen steadily from 16.4 to 24.5 man-rems per megawatt-year. Thus the exposure level at TAPS is a whole order of magnitude higher than that in the USA. (Incidentally, radiation exposures at Magnox nuclear reactors in the UK have been even lower than those at American LWRs.)

For Dresden-1 in particular, radiation exposures seem also to have been considerably below those at TAPS. Over the years 1960–70 exposure levels at Dresden-1 were of the order of 1.7 man-rems per megawatt-year. And in 1976 permanent workers at Dresden-1 received 887 rems, while temporary workers received 729 rems. Exposures are not likely to be any higher in future years, since a radiation decontamination programme was due to get underway by 1979.

Radiation exposures at RAPS-1 are shown in Table 4.10.

Average exposures are much lower here than at TAPS and also there have only been a few cases of exposures greater than 5 rems. However, the figures for man-rems per megawatt-year show that exposures have been high in rela-

Table 4.10 *External radiation exposures to workers at RAPS-1*

Year	No. of workers exposed	Total dose (man-rems)	Average exposure (rems)	Man-rems per megawatt-year
1973	400	162	0.405	4.7
1974	414	364	0.879	5.0
1975	1065	761	0.715	11.5
1976	976	788	0.807	8.9
1977	627	279	0.445	5.3
1978	1018	622	0.611	33.8

Sources: BARC (*Summary of personnel radiation exposures* ... , 1974, 1976, 1977); DAE (1977); *Bulletin of Radiation Protection (India)*, April/June 1979, p. 3.

tion to the amount of electricity produced. If RAPS-1 operates at higher capacity factors in subsequent years, it is likely that radiation exposures will increase also. (The figures for man-rems per megawatt-year vary widely since most exposures occur during maintenance and much maintenance is carried out with the reactor shut down. Therefore a year in which a lot of maintenance work has been done will register high man-rems but low electrical output.

Experience with CANDUs in Canada has shown that internal contamination with tritium represents 25–30% of the total radiation exposure. The tabulated doses above should be increased accordingly to give a true picture.

Table 4.11 shows annual radiation exposures at Douglas Point.

For 1969–71 (the only years for which calculations can validly be carried out) the number of man-rems per megawatt-year was 17.7. This compares very poorly with LWR and Magnox performance. However, between October 1971 and April 1972 the cooling system was flushed out and decontaminated, much reducing radiation levels. The same operation was carried out again in August 1975. It is possible that RAPS-1 radiation levels will be of a similar order of magnitude to those in the early years of Douglas Point, unless similar measures are taken.

To summarise all these statistics, one can say that TAPS

Table 4.11   *Total radiation exposures for workers at Douglas Point*

Year	Internal (rems)	External (rems)
1967	4	57
1968	79	355
1969	220	803
1970	245	1061
1971	393	1542
1972	167	1207
1973	332	891
1974	142	413
1975	192	447

Source: Lesurf (1977).

has by far the worst radiation exposure levels in the whole of the Indian DAE. These high exposure levels give grounds for concern. No deaths have so far been attributed to radiation exposure, but is is to be expected that in fifteen to twenty-five years' time a significant number of cancer-related deaths will occur.

Several reasons can be suggested to explain the occurrence of these high exposures. The reasons range from the very general to the very specific, and some are listed as follows:

(a) Religious thought in India centres around the philosophy of rebirth. Such a philosophy may place less emphasis on the value of a particular life and so lead to a degree of casualness in the face of danger and death.

(b) In a country with a low GNP per capita the economic loss resulting from the average injury or death is likely to be evaluated as correspondingly low. A parallel argument to this is that when people are dying *now* of disease and malnutrition, cancer-related deaths in twenty years' time seem almost unimportant.

(c) There is no lack of highly trained graduate labour and unskilled labour in India, but there is a shortage of intermediate, skilled workers. At TAPS this shortage may sometimes result in welders, say, being given high exposures because there is no one else to do the job at hand.

(d) Filter sludges and spent ion-exchange resins are put through a centrifuge to separate out the solids. These solids are then loaded into 55 gallon drums for storing in underground trenches. Unlike some other BWRs, TAPS has not had a long term storage facility to allow the radioactivity to decrease before carrying out these operations. Since the drumming procedure involves handling the drums, high exposures have resulted. In recent years a storage vault has been constructed to which the spent resins will be pumped in the form of a slurry. This should ameliorate the situation considerably. (A similar manual drumming procedure for spent resin disposal has been a major source of radiation exposure at RAPS-1 also.)

(e) TAPS was supposed to demonstrate that nuclear power in India could be economic. There is therefore a lot of pressure to keep TAPS 'on line' and supplying power. A production-related bonus payment has been introduced in recent years to motivate the workers and improve the station performance. Such a situation will almost inevitably lead to short cuts and risk-taking.

Heavy water losses from RAPS-1

The heavy water in RAPS-1 is contained in two separate systems, the high pressure cooling system and the low pressure moderator system. The bulk of the heavy water (more than two-thirds) is contained in the moderator system. For moderating purposes the heavy water must be of 99.75% isotopic purity or greater, while for the coolant it needs to be of 95% isotopic purity or greater.

The escape of heavy water has been a continuing problem with RAPS-1. This occurs overwhelmingly from the high pressure cooling system, escapes from the low pressure moderator system being negligible. A large proportion of the heavy water that escapes is recovered, but it is mixed with from 20 to 70% of light water from the atmosphere. It is therefore necessary to upgrade it to sufficient isotopic purity for re-use. Table 4.12 shows the relevant statistics for RAPS-1.

During the years 1974–6 about 60–70 tonnes of heavy water had to be upgraded each year, while around 20 tonnes was lost completely. The current annual loss is still of the order of 20 tonnes. In early August 1981 a sudden leak

Table 4.12   *Heavy water escapes and losses from RAPS-1*

Year	Escapes (kg h^{-1})	Recoveries (kg h^{-1})	Complete losses (kg h^{-1})
1972	5.9	4.6	1.3
1973	6.6	4.4	2.2
1974	9.4	7.3	2.1
1975	9.8	6.9	2.9
1976 (provisional)	9.4	7.8	1.6

Source: Schiffer (1976, p. 65).

allowed 8 tonnes of heavy water to escape into a shielded vault in the reactor building. Most of this was recovered and put back into the system.

It seems that at Douglas Point insufficient attention was given to this problem at the design stage. Neither the prevention of escapes nor the recovery of heavy water were adequately thought out. In 1970 the heavy water loss rate at Douglas Point was 1.65 kg h^{-1} (similar to RAPS-1), but equipment was being installed to reduce this to about 0.6 kg h^{-1}. The use of bellows-sealed rather than packed-stem valves in the cooling systems at the Pickering and Bruce reactors certainly seems to have reduced overall losses substantially.

At RAPS-1 various measures have been taken to get this problem under control. There has been added provision of heavy water recovery equipment, bellows-sealed valves have been used, and light water losses from the reactor systems have been minimised (to make upgrading easier). But the loss rate still remains high. For MAPS-1 and MAPS-2 the losses are expected to be around 12 and 10 tonnes per year, and for NAPS 8 tonnes per year.

The value of the heavy water that is completely lost, and the cost of upgrading heavy water that is recovered, are both considerable. The price of heavy water in 1978 was about $210 per kilogram in Canada, and the price India has paid the USSR has been about the same. (By 1979 the cost of heavy water in India was in excess of Rs 2000 per kilogram.) The complete loss of 20 tonnes of heavy water per year at RAPS-1 therefore amounts to an added running cost of $4 million per year. Since this figure neglects additional upgrading costs, we can see that heavy water escapes have a very considerable bearing upon the economics of the CANDU type reactor.

Economics of nuclear power versus conventional power

In any economic analysis there are such a large number of variables whose value can be altered according to one's assumptions that there is plenty of scope for totally different results to ensue. One observer may label a project 'economic', while another calls it 'uneconomic'. While accepting that value judgements lie at the very heart of economics, it is felt that this assessment would not be complete without a brief analysis of the economics of nuclear power in India. Almost all previous analyses of this type have been concerned with the future economic *prospects* for nuclear power, whereas we are now in a position to look back and ask what the realities have been. For instance, prior economic assessments have had to make an estimate of what percentage nuclear capacity factors might be, whereas now it is possible simply to use the actual figures.

Of course, even if one does use the best and most reasonable economic data and one ends up by evaluating a project as uneconomic, it is still feasible to argue in favour of that project on other grounds, in terms of long term political strategy, say. Nevertheless, some appreciation of the purely financial aspects of a project is surely a valuable aid in assessing it.

Capital costs
For power stations of all types the cost per kilowatt varies according to two major factors. First, the cost appears to rise over the years simply due to the general inflationary trend. Second, the cost appears to decrease with increasing size of power units involved.

Tarapur Atomic Power Station was completed in 1969 at a final cost of Rs 970 million for the two units capable of producing 190 MWe net each. Thus TAPS cost about Rs 2550 per kilowatt. It is useful to compare this figure with conventional power stations completed at about the same time. If we choose mainly stations which were completed *after* TAPS, then we shall not bias the argument against nuclear because, due to inflation, later completions are liable to cost more than earlier completions. On the other hand, if we can choose conventional stations with fairly large unit sizes, this will avoid biasing the argument toward nuclear, for the Indian nuclear power units were (at the time of their introduction) the largest in the country.

Coal-fired power stations completed in India between 1968 and 1972 included Delhi, Satpura, Bhusawal, Nasik I, Talcher, Neyveli I, Harduaganj III and Chandrapura II.

Table 4.13 Costs of power generation

	TAPS	RAPS-1	Hydro	Coal
Capital cost (rupees per kilowatt)	2550	3650	2270	1995
Fixed costs (rupees per kilowatt) (interest @ 9%; depreciation @ 5%)[a]	357	511	318	279
Fixed costs (paise per kilowatt-hour) (capacity factor = 50%)[b]	8.2	11.7	7.3	6.4
Fuel costs (paise per kilowatt-hour)	3.7[c]	2.0[d]	—	3.4[e]
Heavy water losses (paise per kilowatt-hour)	—	3.7[f]	—	—
Transmission costs (paise per kilowatt-hour)[g]				
Over 800 km	—	—	2	2
Over 1000 km	—	—	2.5	2.5[h]
	11.9	17.4	7.3 at dam	9.8 at pithead
			9.3 at 800 km	11.8 at 800 km
			9.8 at 1000 km	12.3 at 1000 km

[a] These interest and depreciation rates were typical for the mid 1970s. Insurance has not been included in these fixed costs for lack of information; if it were included it would most probably penalise the newest technology, viz. nuclear power. Operation and maintenance costs are assumed in this analysis to be the same for all types of plant, though in practice they are likely to be cheaper for conventional power stations. (Although nuclear power stations are assumed generally to need less labour than coal-fired power stations, in India, as we have seen, the labour force employed has been comparable to that for a coal-fired power station due to the need to limit radiation exposures. Since the workforce at a nuclear power station has a higher average level of skill, and therefore commands higher wages, labour costs at a nuclear power station will, most probably, be higher than at a coal-fired power station (while labour costs at hydro power stations will be lowest of all).)

[b] A capacity factor close to actual lifetime performance has been chosen.

[c] A year's supply of enriched UF_6 (about 18 tonnes) cost Rs 50 million, while fabrication costs amounted to an additional Rs 800 per kilogram of fuel or about Rs 14.4 million per year. (By 1979 the price of enriched UF_6 had risen to very nearly Rs 4 million per tonne.)

[d] This is an estimated minimum value. Two methods of estimation have been used. First, Indian natural uranium is said to be twice as expensive to produce as Canadian uranium which fetched about $44 per kilogram. Since RAPS-1 working at 50% capacity factor needs at least 25 tonnes of fuel per year, we can calculate the unit cost, excluding fabrication costs, as 1.9 paise per kilowatt-hour. (Fabrication costs elsewhere in the world have been of about the same order as uranium costs, therefore this figure may substantially underestimate the real fuel cost.) Second, the capital cost of the fuel cycle components servicing RAPS-1 (i.e. one-eighth of the uranium mine and mill, one-quarter of the natural uranium fuel fabrication plant, and one-twentieth of the zircaloy plant) amount to around Rs 93 million. At a combined interest and depreciation rate of 14% this gives a unit cost of 1.5 paise per kilowatt-hour. Taking the same fractions of the workforce at the mine and mill and nuclear fuel complex as servicing RAPS-1 gives a workforce of 700 devoted to providing fuel. At a wage of Rs 6000 per year the labour costs represent a unit cost of 0.5 paise per kilowatt-hour. Thus the combined capital and labour components of the fuel cycle cost amount to 2.0 paise per kilowatt-hour.

[e] Coal costs about Rs 50 per tonne at the pithead, and the national average of coal used per unit of electricity generated was about 0.7 kilogram in 1970. (A modern coal-fired station would perform better than this.)

[f] Considering complete losses only (at Rs 32 million per year) and using a capacity factor of 50%.

[g] Using 400 kV lines to transmit 400 MW.

[h] A number of previous estimates of the economics of nuclear power have compared nuclear power with coal power using coal transported to a power station near the demand centre rather than transmitting electricity from somewhere reasonably close to the pithead. Transporting coal by rail is somewhat more expensive than the alternative of electricity transmission, but only marginally so. (Coal transported 1000 km almost doubles in price so that the total cost of electricity becomes approximately 12.7 paise per kilowatt-hour as compared to transmitted electricity at 12.3.) Since Indian Railways has had difficulty in carrying sufficient coal to meet demand over recent years, it would be better to plan for electricity transmission.

These range over power units from 50 to 140 MWe and their average cost was Rs 1380 per kilowatt. (These cost figures, and those that follow for other coal-fired and hydro-electric power stations, are taken from the *Times of India directory and yearbook 1978*.)

Hydroelectric power stations completed between 1968 and 1972 included Upper Sileru, Kuttiyadi, Sabarigiri, Sholayar, Upper Bari Doab, Jawahar Sagar, Kodayar, Obra and Yamuna I. These are all first stages of hydro projects, since it would be unfair to base a comparison on second stage projects which sometimes involve merely the placing of turbine-generators in an already extant and expensive dam. These schemes range in power unit size from 11.25 to 60 MWe and their average cost is Rs 1916 per kilowatt.

Similarly, RAPS-1 was completed in 1973 for Rs 730 million. Thus it cost Rs 3650 per kilowatt.

Coal-fired stations completed between 1972 and 1976 included Dhuvaran II, Ukai I, Ennore II, Harduaganj V, Badarpur I, Koradi I and II, and Chandrapur IV and V. These range in size from 100 to 140 MWe and their average cost is Rs 1995 per kilowatt.

Hydroelectric stations completed between 1972 and 1976 included Lower Sileru I, Srisailam I, Salal, Idikki, Chibro and Ramganga. They range in size from 100 to 130 MWe and cost an average of Rs 2270 per kilowatt.

In the analysis which follows we shall use the cost figures for conventional power stations completed in the period 1972–76, rather than the earlier period. Our comparison using these figures should tend to show nuclear power in a favourable light.

Total generating costs

Table 4.13 shows the total generating costs for nuclear, coal and hydro power stations in India around 1975–6. Only the fuel costs for RAPS-1 have been estimated, the rest of the analysis is based on previously published, and predominantly official, figures.

Assumptions favourable to nuclear power have been made in calculating these costs, yet clearly the conventional power sources of hydro and coal are still cheaper than nuclear. Only when electricity from a coal-fired power station has to be transmitted more than 800 km does it become as expensive as that from TAPS. Since Bombay is one of the demand centres most distant from a coalfield (at around 800 km), we can see that TAPS must be only marginally economic, if at all. As TAPS was bought at a bargain price not likely to be repeated, and since the CANDU style of reactor appears to be so capital-expensive as to be economic only with zero fuel costs and zero heavy water losses, it can be surmised that nuclear power is unlikely to be economic in India in the near future. (MAPS and NAPS are projected to cost about Rs 4800 per kilowatt and Rs 7000 per kilowatt respectively, i.e. even more expensive than RAPS-1.)

Table 4.14 shows the actual selling price for nuclear-produced electricity over the years. In the states of Gujarat, Maharashtra and Rajasthan (to which TAPS and RAPS feed their supply), the average price for conventionally produced electricity sold by the electricity boards in the period 1977–9 was between 21 and 25 paise per kilowatt-hour. Since the electricity boards do not make ends meet, it seems fairly clear that nuclear power is being sold at a slightly subsidised rate at RAPS and at a highly subsidised rate (close to generating cost) at TAPS.

Table 4.14 *Sale price of nuclear electricity (paise per kilowatt-hour)*

	TAPS	RAPS
1973/4	6.0	10.38
1975/6	7.71	
1977	13.38	18.21
1979	13.38	22.85

Sources: DAE Performance Budgets 1977/8 and 1979/80; DAE Annual Report 1973/4, p. 122; DAE (1977, p. 52); IAEA (1974b).

CHAPTER 5

Future Possibilities and Alternative Options

The future for nuclear power

As explained in Chapter 3, the Indian nuclear programme has all along been conceived in terms of a three-stage development from thermal reactors using natural uranium, to fast breeder reactors using plutonium, and on to either fast or thermal breeders using uranium 233 produced from thorium. (A breeder reactor produces more fuel than it consumes. It does this by converting 'fertile material' (such as uranium 238 or thorium) into 'fissile material' (such as plutonium 239 or uranium 233). This transmutation occurs through neutron capture and subsequent radioactive decay processes. Breeding plutonium 239 from uranium 238 is only feasible using fast neutrons, i.e. in a *fast* breeder. Breeding uranium 233 from thorium is feasible with either fast or thermal neutrons.) The rationale behind this three-stage plan has been the desire to utilise India's thorium resources. The existence of large deposits of thorium in India has been known since at least the early years of the twentieth century. Historically the major use for thorium has been in gas mantles, in the form of thorium nitrate or thorium oxide. Since thorium is not of major economic interest these days, estimates of available resources are consequently imprecise. Total thorium deposits in India have been estimated at a minimum of 450 000 tonnes. This figure may be compared to India's estimated uranium resources of around 50 000 tonnes. The majority of India's thorium is located in monazite which is found in the beach sands of the south-west. Monazite makes up about 1.5–3.5% of these beach sands,

82

while thorium makes up about 9% of the monazite. Easily extractable resources of thorium amount to 320 000 tonnes. Of other countries only the USA is thought to have as much low cost thorium as this, while Brazil and Canada each have about half this quantity.

As currently estimated, world uranium resources, at 5 million tonnes (OECD/NEA and IAEA, 1979, pp. 18–19), are reckoned to be larger than thorium resources, at 1.3 million tonnes. However, thorium is considerably cheaper than uranium. This is because thorium can be produced fairly plentifully in comparison to world demand as a by-product of other processes; and also because it is not itself a fissile material, making it less valuable accordingly. In the 1960s, when uranium was cheap, thorium was about two-thirds the price of uranium; in the mid 1970s, when uranium was expensive, thorium was about one-third the price of uranium. This cheapness obviously makes thorium an attractive potential provider of fuel for nuclear reactors. However, two questions arise if thorium comes into demand as a fertile material in a widespread way. First, will additional reserves be found so as to make thorium as significant a resource as uranium? Second, will the price remain attractively low?

These questions relate to the world market for thorium. In India it is likely that thorium will continue to be more abundant and cheaper than uranium, and the same may well apply in Brazil and Egypt. But if these are the only nations where this is true, they may be forced to go it alone in developing the use of thorium. (Canada and the USA have shown interest in utilising thorium, but they would be unlikely to share their knowledge with either India or Brazil since these two countries have not signed the NPT.)

There are a number of different options for transmuting natural thorium (thorium 232) into fissile uranium 233. It is equally feasible to use fast or thermal neutrons, therefore the process can be carried out in either fast or thermal reactors. It is also technically possible to use any of the fissile materials, uranium 235, uranium 233, or plutonium 239, as the fuel and source of neutrons. However, uranium 233 can only be produced from thorium and could therefore only be used in a subsequent stage rather than as a direct option. Uranium 235 could only be used in an enriched form, necessitating an enrichment plant which would cost a lot to build and run. Plutonium has better neutron generation characteristics than the other fuels when fissioned by fast neutrons, and so it is regarded as most valuable for use in fast reactors. But, for reasons which will be explained shortly, a fast reactor burning plutonium as fuel to produce additional plutonium from uranium 238, would be more attractive than a reactor fuelled with plutonium and producing uranium 233 from thorium. In particular the conversion ratio, i.e. the amount of fissile material produced relative to the amount of fissile material consumed, would be higher for the former option than for the latter. The higher the conversion ratio, the more resources of fertile material can be transmuted into fissile material. If the conversion ratio is just below one, then the supplies of fissile material (essentially uranium 235) can be almost doubled. But if the conversion ratio is greater than one (i.e. it becomes a breeding ratio), then it is theoretically possible to transmute all the fertile material available (which, in the case of uranium 238 alone, is more than 100 times as abundant as uranium 235).

To explain the considerations which make a higher breeding ratio possible on the ^{238}U–^{239}Pu cycle as opposed to the ^{232}Th–^{233}U cycle it is necessary to examine the two processes in detail, as shown in Figure 5.1. It can be seen that there is a great deal of similarity between the outlines of the two processes. Each requires not only the absorption of a neutron but also two successive beta (β) decays to reach the fissile state. But a difference between the two processes arises with the different behaviour of protactinium 233 and neptunium 239. Any neutron absorption by these materials removes valuable neutrons from the system and also prevents the beta decay to fissile material. These losses are a function of the decay time, and of the propensity to absorb neutrons on the part of ^{233}Pa and ^{239}Np. Since the half-life for ^{233}Pa is 27.4 days compared to 2.3 days for ^{239}Np, and ^{233}Pa absorbs neutrons at a rate 50% higher than ^{239}Np, it can be said that the transmutation of ^{232}Th can only be carried out less efficiently than that for ^{238}U.

A further disadvantage of the thorium cycle is the pro-

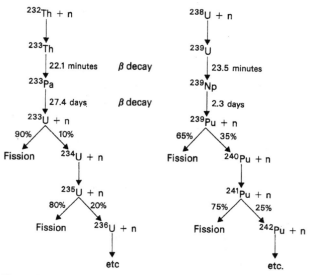

Figure 5.1   Comparison of ^{232}Th–^{233}U and ^{238}U–^{239}Pu cycles.

duction of uranium 232 (not shown in Fig. 5.1). This is produced via a number of routes involving neutron capture and decay by ^{232}Th, ^{233}Th and ^{233}Pa. The ^{232}U in itself is not troublesome, but through a series of alpha and beta decays it gives rise to a long chain of hard gamma emitters, ending in the stable isotope lead 208. The existence of these highly radioactive daughter products complicates the business of reprocessing and refabrication of the fuel. For instance, when reprocessing is carried out to extract ^{233}U, the ^{232}U inevitably comes with it (since reprocessing works by chemical separation and not physical separation). The subsequent decay of this ^{232}U and the resultant radioactivity mean that the fabrication of the ^{233}U into further fuel can only be carried out using remote handling techniques. Similarly, the re-use of non-transmuted ^{232}Th through reprocessing would also involve the acquisition of ^{228}Th, which is the first daughter product of the ^{232}U decay chain. Therefore it is thought that thorium would not be recycled for about 15–20 years, in order to allow the decay of ^{228}Th. These problems do not occur with the ^{238}U cycle since ^{238}Pu decays to ^{234}U, which has a long half-life and is therefore not very radioactive.

There are two advantages that the ^{232}Th cycle has over the ^{238}U cycle. First, breeding in thermal reactors is only possible using ^{233}U as fuel. As we have mentioned, however, this is not an immediate option since the ^{233}U has first to be produced somehow. Second, although the performance of fast breeder reactors would be better with ^{238}U than with ^{232}Th, the use of a mixed uranium–thorium cycle might be considered in the interests of better safety characteristics. For with a mixed uranium–thorium cycle the reactor would tend to have a negative coefficient of reactivity, meaning that under unusual conditions the reactor would tend to shut down. Whereas, with a conventional liquid metal-cooled fast breeder operating on the uranium cycle alone the reactivity tends to rise with, say, an increase in temperature.

In the light of these advantages and disadvantages for the thorium cycle it is possible to understand why India is aiming at plutonium-fuelled fast breeders (similar to those under development elsewhere) for the second stage of the nuclear programme. A specially designed thermal breeder could not be set up without ^{233}U as fuel, and this type of reactor would have to involve very novel techniques such as circulating molten fuel. Furthermore, a thermal breeder would have a lower breeding ratio than a fast breeder reactor. To use thorium in an ordinary but modified thermal reactor (such as the CANDU) would involve fuelling it with plutonium and yet only achieving a conversion ratio of just under one. Therefore, India has chosen to go for plutonium-fuelled fast breeders which, it is intended, could produce plutonium and ^{233}U.

However, it is possible to see that this is not an entirely satisfactory method of developing fissile resources. For, on the one hand, India does not have a uranium enrichment plant and therefore does not have significant supplies of depleted uranium (i.e. uranium high in ^{238}U because the ^{235}U has been largely removed) to use as fertile material in the plutonium–uranium 238 cycle. The alternative is to use the ^{238}U from reprocessed CANDU fuel and incur an economic penalty for handling radioactive material. And, on the other hand, as we have seen, the production of ^{233}U from thorium is not as attractive a proposition as the production of ^{239}Pu from enrichment-depleted natural uranium. The resultant hybrid programme may be in danger of falling between two stools.

India is now beginning to embark on the second stage of her programme. Currently a Fast Breeder Test Reactor (FBTR) is being built as part of a Reactor Research Centre at Kalpakkam near Madras. This reactor is being set up using French assistance. A first agreement between the DAE and the French Atomic Energy Commission was signed in April 1969. Subsequently, several French industrial companies (Groupement Atomique Alsacienne Atlantique, Hispano-Suiza, Creusot-Loire and Stein Industrie) entered into agreement for the supply of materials, equipment and know-how. Construction work began around 1972 and the estimated cost for FBTR was then Rs 345 million, including a foreign exchange component of Rs 80 million. The French government agreed to a special foreign exchange credit equivalent to Rs 47 million, while the rest of the foreign exchange requirement was to be met out of normal economic assistance from the same source. The Reactor Research Centre as a whole (including FBTR) was then estimated to cost Rs 680 million.

It was hoped that FBTR would be complete in about four years, but it is now thought that it may be ready by mid 1982. The cost has escalated considerably, the latest estimate for FBTR being Rs 697.2 million (DAE Performance Budget 1979/80), while the Reactor Research Centre will cost well over Rs 1000 million.

FBTR is essentially a copy of the Rapsodie-Fortissimo reactor which went into operation in France in 1970. The main difference between the two being that the latter produced no electrical power whereas FBTR will do so. The Rapsodie reactor was intended to be a true breeder reactor having a conversion ratio of about 1.1–1.3. But in practice its conversion ratio was very low, around 0.33. It is intended that FBTR will have a conversion ratio of 0.5–0.6; it will not, therefore, be a true breeder.

FBTR will use sodium coolant and its output will be about 42.5 MWth, leading to an electrical output of 12–15 MWe. This is rather small in comparison to other fast breeder reactors which are operating or under construction around the world. It will use mixed oxide fuel consisting of 30%

PuO_2 and 70% UO_2, the uranium having a ^{235}U content of 85%. This highly enriched uranium is to be obtained from France, while the plutonium is to be produced in India by reprocessing fuel from the Indian CANDUs. (The DAE investigated the possibility of using a purely indigenous fuel consisting of plutonium and natural uranium, but this turned out to be an unattractive proposition.) Fabrication of the mixed oxide fuel is to be done at BARC. The PuO_2 content will be about 54 kg, and the UO_2 content about 127 kg. A laboratory-scale reprocessing plant is under construction at Kalpakkam. This will reprocess some of the spent fuel from FBTR, but it is too small to reprocess all the fuel, so that a larger plant would be needed in future.

The 'blanket' surrounding the FBTR core will contain ThO_2 pellets. It is intended that ^{233}U produced in the blanket will be used to replace the ^{235}U in the core as it gets used up. (A pilot plant for the separation of ^{233}U from irradiated thorium was successfully operated in 1970. It was set up in the reprocessing plant at BARC and used thorium irradiated in the CIRUS reactor.)

Hope has been expressed lately that a fast breeder power reactor with plutonium fuel and a thorium blanket will be in operation in India by 1990. Indeed, in 1976 the DAE mentioned a target of 14 000 MWe of fast breeders by the year 2000. Both of these targets seem unattainable. In order to provide a fuel inventory for 14 000 MWe of fast breeders it would be necessary to acquire about 56 tonnes of fissile plutonium. It would take 8000 MWe of CANDU reactors operating at a high capacity factor to produce this quantity over a period of 20 years. Since at present only 400 MWe of CANDU is actually operating, and only a further 900 MWe or so is under construction, it will certainly not be possible to attain the target. The picture is not at all altered by considering past and future plutonium production from Tarapur since its plutonium output is low (about 80 kg fissile per year). Neither would plutonium production from the fast breeders themselves make a significant difference, since each breeder is likely to take at least twenty years (and possibly considerably more) to produce enough fuel to start another identical breeder reactor.

In considering the scope of a future fast breeder programme it is necessary also to remember that plutonium tends to depreciate with extended storage. This is because the plutonium produced in reactors consists not only of plutonium 239 but also ^{240}Pu, ^{241}Pu and ^{242}Pu, which are formed by successive neutron capture. Now the ^{239}Pu and ^{241}Pu are fissile, whereas the ^{240}Pu and ^{242}Pu are not. The ^{241}Pu has a half life of only 14 years and it decays to americium 241, which is non-fissile and highly radioactive. Thus storage means that the fissile content is reduced, and that there will be additional handling problems, perhaps necessitating chemical separation.

It appears unlikely that thorium could become a significantly useful resource in India before the year 2000. Consequently nuclear power in general is likely to be insignificant before that date.

Alternatives to nuclear power

Since nuclear power has not yet become a major energy source in India, it is not hard to find alternatives to the present nuclear capacity. Two or three coal-fired or hydro-electric power stations could have substituted quite easily. However, as far as the short term and the long term future goes, it is interesting to ask what range of options may be open. In answering this question one may consider conventional as well as non-conventional (usually renewable) energy sources. For the short term the conventional sources offer the most reliable option. Coal, oil and hydroelectricity have all proved their worth over many decades. In India, coal and hydro potential are likely to be in ample supply – the latter until the year 2000 and the former until a long time after that. And it is likely that further large quantities of oil will be found. The only physical difficulty with coal and hydroelectric energy seems not to lie with their abundance but with their location. There appear to be several ways of circumventing this difficulty. First, greater effort could be expended on detailed surveys for coal and hydro potential in areas not high in energy sources. Second, the regional electricity grids could be strengthened and linked into a national grid so as to transfer energy more easily from place to place. Third, more R & D effort could be devoted to dispersed and decentralised renewable energy sources which could offer not only a supplement to conventional energy but also perhaps a much more viable source for the rural areas.

A decision to retain coal as a major energy source need not imply a major problem from atmospheric pollution. Coal gasification, as suggested by S. H. Zaheer, might well be applicable in certain areas. Even if coal is burnt to produce power, modern techniques such as 'scrubbing' devices and fluidised bed combustion can minimise the environmental effects. Such techniques would certainly raise the capital cost of a coal-fired power station, but probably not to the level of a nuclear power station.

Current thinking among India's planners envisages substantial increases in conventional generating capacity over the next few years, with hydropower to grow by 50% (from 10 to 15 GW) and coal power by 100% (from 13 to 26 GW). Later, by the end of the 1980s, it is planned that there will be five Super Thermal Power Stations operating, fuelled by coal from nearby collieries. These are to have a total capacity of 7.2 GW, based on unit sizes of 200 and 500 MWe, and are to be interconnected by a 400 kV transmission system. (The five stations are to be widely spread about the country, at Singrauli, Uttar Pradesh; Korba, Madhya Pradesh; Ramagundam, Andhra Pradesh; Neyveli, Tamil Nadu; and Farakka, Bihar.) All of them will have high stacks and electrostatic precipitators for pollution control.

So for the next decade at least, coal and hydro will be the main power sources in India. In the longer term it is possible that renewable energy sources could provide a significant proportion of India's needs.

In common with most other countries of the world, India neglected study of renewable energy sources while oil seemed inexpensive. During the 1960s and early 1970s research and development work on such topics as solar and wind energy was not fashionable (although it had been in previous years). But by the late 1970s several government establishments and large private companies were involved in, for instance, solar energy research. These organisations included Bharat Heavy Electricals Ltd, Central Electronics

Ltd (Government of India), Tata, Kirloskar, Jyoti, Amul, Mahendra and Metal Box Co.

This is not an appropriate place to attempt an extended review of work on renewable energy sources in India. The subject area has too broad a scope to be covered effectively in just a few pages. However, a word or two on the potential for some broad areas may not come amiss.

Solar energy certainly has great potential in India. For about 75% of the year sunshine throughout the day is assured for most of the country. During the monsoon, cloud cover makes direct sunlight an unreliable source but the diffuse sunlight available may well be sufficiently powerful to be worth using. An indirect means of obtaining solar energy which is already used on a significant scale is that of biogas production. Robin Roy (1980) estimates that there were 80 000 biogas plants at work in India in 1979, most of them of a small size, but sufficient to supply the needs of a family. However, the Chinese have used biogas on a far more extensive scale. Roy estimates that there were 7.2 million biogas plants in China in 1979. The Chinese plants use pig manure whereas the Indian plants use cow dung. (Pigs being ubiquitous in China, and cows ubiquitous in India.) Pig manure seems to be the more suitable material for biogas production, which might explain why the Chinese programme is much bigger than the Indian. Thus some kind of biochemical breakthrough in methods of handling source material for Indian biogas plants might be necessary to bring the Indian programme up to the size of the Chinese. When biogas production has ceased, a nitrogen-rich material is left behind. This is suitable for use as a fertiliser. Since the manufacture of artificial fertilisers is an energy-intensive activity, this 'by-product' of biogas production may represent a substantial way of saving energy. A biogas programme of Chinese dimensions (seven million family-size plants) would produce around 5 million tonnes of nitrogen per year. This may be compared to India's annual consumption of nitrogenous fertilisers of around 2.5 million tonnes in the early 1970s. (About two-thirds of this consumption was imported.)

Another indirect means of capturing solar energy is wind power. However, its use on a substantial scale in India does not seem likely. Winds of a reasonable speed only occur around the time of the monsoon. Wind energy might then be used for water pumping, but investment in the production of electricity does not seem attractive.

Solar energy, obtainable in a multitude of ways in the form of heat, electricity (solar cells), or chemical energy, represents an immense potential for India. However, two energy sources of non-solar origin could also be used – tidal and geothermal energy. Both of these sources have the disadvantage, though, of being site specific. Geothermal energy is most easily available in the north of India in the seismic zone along the Himalayas. Since the hydroelectric potential in that area is great, the production of geothermal electricity would have to be economically competitive with hydro if it were to be developed on a substantial scale. It is doubtful if geothermal power could be competitive, although geothermal *heat* might find some local use on a small scale.

The main potential sites for tidal power are the Gulfs of Kutch and Cambay on the west coast. Depending upon the size of barrage built, the capacity of these two schemes could be up to several thousand megawatts. However, the maximum tidal energy is available twice per day at times which may not correspond with energy demand. Thus such schemes would operate at quite low capacity factors, say 20%, producing an actual energy output of several thousand GWh. E. M. Wilson, of Salford University, estimated the cost of these schemes (Wilson, 1977), the details were as follows:

Gulf of Cambay	7364 MWe	Rs 19 250 million
Gulf of Kutch	1187 MWe	Rs 5940 million

In terms of capital cost per kilowatt, then, these tidal schemes would be roughly comparable to nuclear power stations, but their energy output (their capacity factor) would be one-half that for nuclear. However, the fuel costs of tidal schemes would, obviously, be zero.

CHAPTER 6

Conclusions

This final chapter will summarise the findings on India, and extend it into an analysis of nuclear power in the Third World to cover the prospects for its future.

India

The range of energy options open to India is broad. The conventional energy sources available (coal, hydroelectric power, oil, natural gas) are under-explored and under-utilised. The non-conventional renewable energy sources are under-researched.

There is an insufficient electrical capacity to meet peak demands. At least three responses can be made to this situation, two of them technological and one economic. First, one could provide extra electrical supply capacity. Second, one could strengthen the transmission network so as to meet peak demands by inter-regional power-sharing. Third, one could reduce demand somewhat (or at least reduce the growth in demand) by raising the price of electricity, at the same time providing extra finance for investment in power supply (as long as demand is not depressed so much as to provide only the same total revenue as before).

The major emphasis in India seems to have been on the first response – to provide additional supply – and nuclear power has been chosen to provide part of that additional supply. The performance of the nuclear power programme leaves a lot to be desired. The growth of nuclear capacity has been very slow, there has been a long gestation period leading up to the operation of commercial reactors, and most individual reactors (and also the heavy water plants)

93

have suffered from fairly considerable delays in their construction. The performance of the completed reactors is not very good. Their actual output as compared to their possible maximum output is about the same as for coal-fired and hydroelectric power stations in India (around 45%). The high capital costs of nuclear reactors dictate that they must be run at something like 70% or more of maximum output in order to be economic. Thus the failure to achieve a better output than other power stations in India indicates one reason why nuclear reactors have not, in actuality, been economic producers of electricity.

In two other respects, operating experience with nuclear reactors has been unsatisfactory in India. At the Rajasthan reactor heavy water losses have been a continuing financial drain, and at Tarapur radiation exposures to workers at the plant have been far too high in comparison to internationally accepted practice. The consequences of the latter situation will only show up in about twenty years' time.

Despite a major effort in uranium exploration (absorbing enough finances to have built one of the Tarapur reactors), India has still not located, after thirty years, any reserves of good quality uranium. The estimated total quantity of uranium available (assured and probable) amounts to the energy equivalent of less than 1% of India's coal reserves. A really sizeable nuclear power programme could not be fuelled by the limited quantity of assured reserves.

India has strived for self-reliance in nuclear power and has made a massive financial investment in nuclear fuel facilities as well as reactors. Twenty to twenty-five per cent of the country's research and development spending has gone on nuclear research. Nevertheless, nuclear power is not yet a major energy source in India, and self-reliance has not yet been achieved. Without an enrichment plant India is still dependent on the USA for fuel supplies for the Tarapur reactor. And until the technical problems with the novel heavy water plants are solved, heavy water too will have to be imported. It is likely that India *can* become self-reliant in the end – the knowledge and capability are there – but at what economic cost? India might have eschewed self-reliance and invested purely in nuclear power stations without much supporting infrastructure (i.e. no reprocessing plant, no heavy water plants, no fuel fabrication, etc). This way the capital investment in actual energy generation could have been higher, and thus more immediately productive. But would India have been capable of operating nuclear power stations without the background knowledge and experience that comes from operating (and researching into) the associated fuel cycle services? And in the long run would it have been sensible to remain dependent on foreign suppliers? Without a reprocessing plant, India could not have employed a nuclear device and so would not have caused an international furore leading to problems with nuclear collaborators. But might some other international issue have provoked a similar degree of sanction (strict in the case of Canada, and partial in the case of the USA)? Such sanctions would have been more crippling to a less self-reliant programme.

The progress made toward self-reliance to date is adequate testimony to the high capability of Indian scientists and engineers. The poor performance of India's reactors may be put down to (a) the acquisition of designs of an experimental, early type; (b) a lack of time (as yet) for learning about their detailed idiosyncrasies of operation; (c) too small a grid size, in the beginning, to cope with a nuclear reactor.

It was probably a mistake to have attempted to build up the nuclear programme without gaining experience upon a prototype power reactor. The move straight into commercial reactors has resulted in problems with radiation exposures and heavy water losses, either of which might have been avoided if experience had been gained on a prototype. Of course, the fact that the problem with radiation exposures is essentially only with the BWR, and that the problem with heavy water losses is only with the CANDU, would imply the need to have built two prototypes, at double the expense. But that indicates a fault in the decision to choose two reactor types, rather than with the policy of building a prototype first.

As things stand, the Rajasthan reactor is essentially a demonstration reactor rather than a fully fledged commercial one. Its overall performance may improve with time due to design modifications and better operating conditions,

but it is probably unlikely that it will approach the performance of, say, the Pickering 600 MWe units in Canada. It will
be several years before India considers the possibility of
building 500 MWe reactors, and meanwhile it is unlikely
that a decision will be taken to invest in a major redesign
effort on the 200 MWe reactor. So only piecemeal design
changes will be made, and therefore only piecemeal
improvements.

It has been shown that a detailed economic analysis of
India's power reactors (on pages 76–81) indicates that
nuclear electricity generation has no advantage over hydro
or coal-fired generation. Indeed the latter two are considerably cheaper unless the electricity must be transmitted
800 km or more.

Taking into account all the reasons enumerated here, and
some more reasons explained in the body of the text, it
appears that nuclear power will not be a significant source of
electricity in India for at least the next twenty years. Who
knows what will happen in the energy scene over the next
twenty years? It could be that completely novel energy
technologies will be developed, or that a radically new type
of nuclear reactor could be designed, superseding present
types. In his presidential address at the first conference on
the Peaceful Uses of Atomic Energy, Homi Bhabha predicted that fusion energy would be available within two
decades. In this he was wrong, but those two decades were a
time of inexpensive and plentiful energy when there was
little incentive to develop new sources. After the oil price
rise of 1973–4 there has been much greater activity in all
forms of energy development. The possibility of new
departures in energy technology and supply is thus much
greater. Would it perhaps be most sensible for India to
diversify her R & D, rather than concentrating on one energy
source for such a very long period of time?

It has been clear that one motivation behind India's
nuclear power programme has been the desire to stay
abreast of modern developments in science and technology.
Yet this might more surely have been achieved by spreading
funding across a number of different scientific areas and
disciplines. Granted that nuclear energy has quite a number
of different aspects – it requires work to be done in several

fields of expertise and has multiple uses (electricity production, isotopes, weapons) – nevertheless there are very many areas of science and technology that are not in any way connected with nuclear energy and which could benefit from extra research and development.

Comparison with other Third World programmes

There are three major energy-related points to take account of when considering embarking on a nuclear power programme (these are not the only points but they seem the most vital):

(a) Does a paucity of other energy supplies make it logical to develop nuclear power?
(b) Is the existent electrical capacity sufficient (if that capacity is thoroughly interconnected) to cope with a reactor of economic size (that is, about 6000 MWe capacity or greater, for a 600 MWe reactor)?
(c) Are there proven indigenous uranium reserves of a reasonable quantity (that is, greater than 6000 tonnes at an absolute minimum; 6000 tonnes is only sufficient to fuel two 600 MWe LWRs for a lifetime of around twenty years)?

Table 6.1 indicates a rough evaluation of where the countries surveyed at the end of this chapter stand with respect to these considerations. Those with a *prima facie* case for nuclear power are Argentina, Brazil, China, Mexico, South Africa and Taiwan. For none of these, though, are all three considerations positive. The other countries are marginal cases, except for Chile, Iraq, Pakistan and the Philippines, where nuclear power does not seem at all a logical development. The existence of military regimes in Chile and Pakistan make one suspect that they may be interested in nuclear weapons rather than nuclear power. Indeed, a primary interest in weapons (resulting from India's acquisition of nuclear capability) is almost certain in the case of Pakistan.

Now it must be admitted that this analysis and classifi-

Table 6.1 *Nuclear power programme evaluation*

Country	Electrical capacity >6000 MW	Energy supplies limited	Uranium reserves >6000 tonnes
Argentina	√	×	√
Brazil	√	×	√
Chile	×	×	×
China	√	×	√
Cuba	×	√	×
Egypt	×	○	×
Iran	○	×	?
Iraq	×	×	×
Israel	×	√	×
Mexico	√	×	√
Pakistan	×	×	?
Philippines	×	×	×
South Africa	√	×	√
South Korea	√	×	×
Taiwan	√	√	×
Turkey	○	√	×

√ = yes; ○ = marginal; × = no; ? = not known.
Source: This chapter, *passim.*

cation is somewhat gross and simplistic. This can be seen clearly if we consider how India would fare in this scheme of analysis. Electrical capacity and uranium reserves in India, it would appear at first sight, are both sufficient to warrant a positive 'yes' to nuclear power. But in fact, electrical capacity is only interconnected in four regional grids rather than a national grid, so that each grid size (c. 6000 MWe) is only just big enough for one reactor. And the uranium India has is low grade and expensive, making this consideration far more marginal too. It is likely that, in the case of the other Third World countries, a finer judgement would also result in more marginal cases.

Having seen that, on energy considerations, nuclear power does not seem entirely logical or necessary in any of the Third World countries surveyed, the question arises as to why these countries have opted for nuclear power. Has it been because the nuclear industries of the USA and Europe have foisted reactors upon ignorant nations? Although nuclear manufacturers have been eager to sell reactors to the Third World to open up new markets and to absorb idle manufacturing capacity, this hardly seems an adequate explanation. Most of the nations involved established atomic energy commissions and nuclear training programmes long ago in the 1950s. Thus both sides have been eager to make nuclear transactions. But whether either side has gathered any economic benefits from these deals is doubtful. As noted in Chapter 1, the American and Canadian nuclear manufacturing industries have not been very profitable, even without considering whether there may have been opportunity costs to the state in giving concessionary financing through export credit agencies. As for the purchaser countries, Table 6.2 shows the most reliable estimates for the capital cost of those reactors in the Third World that are fully completed. The sample is a small one, and is likely to be unrepresentative since the prices for these first sales have been deliberately kept low. However, we can see that the price for the Indian contracts in dollars per kilowatt compares quite well with the others, i.e. the Indian reactor costs are quite low. Given that the economics of these Indian reactors (as analysed in Chapter 4) is very poor, then the economics of the other reactors named is likely to be very poor also.

It has not been possible to carry out a detailed analysis of the financial expenditure incurred by those Third World countries included in this survey. Therefore, the interesting question of whether countries other than India have spent such a large proportion of their R & D budget on nuclear energy must be left to other researchers.

Neither has it been possible to find out whether (as has not occurred in India) the growth of nuclear capacity has met the expectations of the nuclear Third World. It may well be too early to say whether nuclear projects have fallen significantly behind schedule or not. Such a review should probably be done in the late 1980s.

How do the reactors in the rest of the Third World compare to India's for operating experience? Data on capacity factors is shown in Table 6.3. It can be seen that only Argentina's reactor has a good performance, despite the fact that due to lack of demand it is operated at a reduced power level over weekends. The rest of the reactors surveyed give a rather mediocre showing. KANUPP in particular has

Table 6.2 Capital costs of Third World reactors

Country	Reactor	Size (MWe)	Capital cost ($ million)	$ per kilowatt	Date of cost estimate
Argentina	Atucha-1	320	145	453	1974 (complete)
Pakistan	Kanupp	125	60	480	1966 (fixed price)
South Korea	Kori-1	590	320	542	1977 (complete)
Taiwan	Chinshan 1 + 2	1200	750	625	?
India	TAPS 1 + 2	380	120	316	1974 (complete plus modifications)
India	RAPS-1	200	90	450	1974 (complete)

Sources: Duayer de Souza (1979, p. 215); Graham and Stevens (1974); Richardson (1978, p. 96); DAE Performance Budget 1977/8, p. 36.

Table 6.3 *Reactor capacity factors*

Name	Type	Country	Size (MWe)	Capacity factor over period
Atucha-1	CANDU	Argentina	320	75.4%, 1974–80
KANUPP	CANDU	Pakistan	125	37.6%, 1973–8
Kori-1	PWR	South Korea	590	50.0%, 1978–80
Chinshan-1	BWR	Taiwan	600	52.4%, 1977–February 1981
Chinshan-2	BWR	Taiwan	600	63.0%, 1979–80
TAPS-1	BWR	India	200	48.2%, 1969–80
TAPS-2	BWR	India	200	51.7%, 1969–80
RAPS-1	CANDU	India	220	40.6%, 1973–80

Sources: IAEA (*Operating experience* ∴, 1970–80); *Nuclear Engineering International*, March 1981; *Nucleonics Week*, 26 March 1981.

Table 6.4 *Total man-rem exposures*

	Atucha-1	KANUPP	Kori-1
1976	236	243	–
1977	1051	94	–
1978	477	136	192

Source: IAEA (*Operating experience* . . . , 1977–80).

suffered from many problems, including transmission line failures and operator errors, all of which have led to a considerable cumulative period of outage. KANUPP has also been plagued by heavy water leaks, though Atucha-1 seems not to have had such severe problems, its heavy water losses being 2.6 tonnes in 1977 and 2.1 tonnes in 1978. It would be interesting to know whether the Chinshan reactors, being BWRs, have, or are acquiring, the same radiation exposure problems as TAPS, but no data on this have been found. Total man-rem exposures at Atucha-1, Kori-1 and KANUPP have been as shown in Table 6.4. These do not, *prima facie*, look as serious as those at TAPS, but more details of the number of persons exposed and their maximum dose would be needed to make a full comparison.

Prospects for the Third World

The IAEA *Market survey for nuclear power in developing countries*, produced in 1974, has been mentioned already. For India it said that 4.2 GW of nuclear capacity was possible by 1980 and that the following additions to capacity (in MWe) could be contemplated:

1981	1982	1983	1984	1985
2 × 800	2 × 800	2 × 1000	2 × 1000	2 × 1000

1986	1987	1988	1989	1990
2 × 1200	3 × 1200	3 × 1200	3 × 1200	4 × 1200

Thus the total installed capacity of nuclear power could be 31.4 GW by 1990 and 130 GW by 2000.

These figures look absurdly optimistic now, given an understanding of India's nuclear development over the last twenty years. But it is not just the figures for India that are optimistic, the prospects for the rest of the Third World are also overstated. The IAEA *Market survey* overestimates the attractiveness of nuclear power to the Third World because, first, it assumes that small sizes of reactor (i.e. 150–400 MWe) would be made commercially available. The only small scale reactor now available is a 125 MWe reactor from Technicatome of France, and this size is generally reckoned to be too small to be economic. Technicatome is now working on a design for a 300 MWe PWR for sale to the Third World. Second, it assumes that the capital costs of reactors in underdeveloped countries would be about 20% less than those in the USA. Third, it ignores energy resources indigenous to the Third World such as coal and natural gas – that there would be some element of competition between these is undeniable. Fourth, the survey was carried out before the formation of a uranium cartel and the consequent very substantial rises in the price of uranium. All these factors combine to mean that the *Market survey* was a purely hypothetical exercise.

It has been indicated in this work that the economics of nuclear power in the Third World is poor, for vendor and purchaser. If this view becomes generally accepted worldwide, then the sale of nuclear reactors will be limited. There would still be some trade in reactors, but purchases would be made almost entirely for weapons-related purposes. Each purchaser country would require only one or two small reactors to provide sufficient plutonium for deterrent effect. In this case one of the major constraints upon the spread of nuclear reactors to the Third World would be the vendor countries' attitudes to nuclear proliferation. The USA apparently became stricter on this issue under the Carter Administration, but in actual fact it did not adopt a hard line position with India. However, given the kind of pressures applied to India, one might judge it possible that Third World countries will decide that it is not worth developing nuclear power *unless* they also want to develop weapons potential. If a country is to be regarded as suspect and unworthy of co-operation purely because it has a strictly peaceful nuclear power programme, then one may as well aim for military use from the very beginning.

The Nuclear Supplier's Group (the so-called London Club) seems to have been fairly ineffectual so far in terms of limiting proliferation. This Group had its origins in the Nuclear Exporter's Committee (or Zanger Committee) formed in 1970 to implement the Non-Proliferation Treaty. The members of the Nuclear Exporter's Committee were Australia, Canada, Belgium, Finland, West Germany, Italy, Holland, Norway, Sweden, Switzerland, the USA, the USSR, the UK, Denmark and Japan. The Nuclear Suppliers' Group was formed in late 1974, partly, it appears, in response to the Indian nuclear explosion. The initial members of the Group were Canada, France, Japan, the UK, the USA, the USSR and West Germany. Additional members were Belgium, Czechoslavakia, East Germany, Italy, Holland, Poland, Sweden and Switzerland. Thus by 1978 the number of participants was fifteen and covered all the nations who were at that time capable of exporting nuclear equipment. (India was invited to join the Group but declined on the grounds that it was a 'northern cartel'.) Broad agreement was reached on a set of guidelines for the export of 'sensitive' nuclear equipment and materials in September 1977, whereby an importing nation must give assurances that such imports would not be used to manufacture nuclear explosives and would be subject to IAEA safeguards. (The technologies of fuel enrichment, reprocessing and heavy water production were designated as 'sensitive'. The guidelines agreement is reproduced in SIPRI (1978, pp. 138–44).) Furthermore, no such imports should be transferred to a third nation without agreement of the supplier country, and any replicated facilities would be subject to IAEA safeguards. It had earlier been proposed that any importing nation would have to accept full scope safeguards over all aspects of its nuclear programme to be eligible for co-operation with Group members, but this was opposed by France and West Germany. The guidelines also stipulate that if a member country has a dispute over violation of safeguards with an importing nation, then all member countries will restrict nuclear assistance to that nation. The reasons why the Group has been ineffectual are probably most strongly related to the difficulty of formulating any technological or political means of limiting proliferation short of banning all nuclear exports. The Nuclear Suppliers' Group's primary aim is, of course, to export nuclear technology, and only secondarily to limit proliferation. Thus one of the main activities of the Group has been to try and agree on suitable safeguards agreements for all contracts so that competition for sales would be on technical and economic considerations rather than on weakness of safeguards.

The IAEA seems not to have taken a very strong line so far on the issues of proliferation and export of 'sensitive' technologies. The IAEA Board of Governors approved the West German–Brazilian package deal and the sale of reprocessing plants by France to Pakistan and South Korea. And despite pressure from the Nuclear Suppliers' Group, it has not adopted similar 'guidelines' as part of its policy, though it continues of course to implement safeguards checks.

If economic considerations are overridden by Third World nations and they go ahead with substantial nuclear power programmes, they are likely to find, in the end, that its contribution to development has been small. Electricity in general, and nuclear power in particular, is not a resource of much interest to the impoverished city dweller or to the rural masses since it is sold at a price far beyond their means and requires all kinds of capital investment on their part to be actually useful (which may appear a marginal expenditure to the rich Westerner but is a major expense to them). On the other hand, a resource such as paraffin has a very wide market and can be used in cheap and simple stoves, etc. It would therefore make far more sense to invest in oil exploration and production than in nuclear power. The prospects for oil industry development in the Third World remain very good. Large areas remain unsurveyed and unexplored, even though the geological conditions are known to be suitable for oil deposits. In the thirty year period since World War II the success rate for exploratory drilling in the Third World has been nine times higher than that elsewhere (Grossling, 1979). Since 1979 the World Bank has been able to finance exploration and production of oil (a move that is to be welcomed) and the prospects for the Third World seem bright (IBRD, 1979). The development of oil in the Third World has long been held back by the multinational oil companies, but the advent of OPEC has changed the political constitution of the world oil industry, leaving far more scope for the Third World.

It is difficult (in fact, perhaps impossible) to forecast the future of energy demand in the industrialised countries, even though these are the countries where most researchers are based and on which most research is done. How much more difficult is it then to forecast energy demand in the Third World, and to plan for an appropriate supply? In particular, is it justified to project future Third World energy requirements on the basis of the rich countries' energy demands at the present time? On the one hand, on the grounds of fairness and justice it might seem essential to allow under-developed countries the freedom to rise to the rich countries' level of energy use. But on the other hand, perhaps our level of energy use is simply wasteful, and further to that, perhaps we could not afford to use energy the way we do if we did not extract resources from the Third World to pay for it. On this latter view, the prospect of the whole world using energy as the developed countries do now is a mirage.

Nuclear power in the Third World: a survey

Having carried out a detailed assessment of India's nuclear power programme, it will be enlightening to survey nuclear developments in other Third World countries so as to gauge how typical (or otherwise) the Indian programme is. So this section will provide thumbnail sketches of the history and current situation of other countries' nuclear programmes.

Each country-based sketch will begin with a brief outline of the present energy supply situation, continue with an account of nuclear developments to date (including the exploration for and discovery of uranium resources) and end with a note on that country's policy with respect to nuclear weapons. South Africa and the People's Republic of China have been included in this survey because, although by some definitions they might not be included in the term 'Third World', they are, nevertheless, very interesting cases. Essentially, only those countries that have nuclear power (or will do so fairly soon) have been included. The following countries are not surveyed, but have expressed an interest, sometimes a very strong interest, in acquiring nuclear power: Algeria, Bangladesh, Colombia, Ecuador, Greece, Indonesia, Jamaica, Jordan, Kuwait, Libya, Malaysia (with Singapore), Morocco, Panama, Peru, Saudi Arabia, Sri Lanka, Syria, Thailand, Uruguay and Venezuela.

Several of these countries have research reactors, e.g. Colombia, Indonesia, Libya, Malaysia, Thailand and Venezuela. Indeed several countries not mentioned in this chapter at all (e.g. Ghana) have research reactors, but have expressed no desire to acquire a commercial power reactor. Of the above-mentioned list it seems that Libya may possibly get a Russian-style PWR, probably of 440 MWe and at an estimated cost of $330 million.

Argentina
Energy supply statistics for Argentina are approximately as follows: oil, 65%; natural gas, 25%; hydroelectric, 10%.

Only 10–15% of this oil consumption is imported, and the majority of the natural gas is indigenous. Investment of $10 000 million is envisaged, leading to self-sufficiency in oil by the later 1980s. Proved reserves of oil amount to 2500 million barrels, and of natural gas to 8000 billion cubic feet. A gas pipeline network covers the length and breadth of the country. There are small reserves of coal, around 500–700 million tonnes, but these are not utilised to any great extent, production being around half a million tonnes per year.

Total electric generating capacity is about 10 000 MW. A national grid system is in operation. Hydroelectric power makes up around 2500 MW of capacity. Total hydroelectric potential has been estimated at 30 000–50 000 MW.

The National Atomic Energy Commission (CNEA) was set up in 1950. Argentina's first research reactor, purchased from the USA (RA-1), went operational in 1958 (Sabato, 1973). Two more research reactors, RA-2 and RA-3

(8 MW) were built in the 1960s. These latter were essentially of indigenous construction, though all the research reactors used highly enriched uranium (imported from the USA) for fuel. (Due to difficulties over the US Nuclear Non-Proliferation Act, further supplies of enriched fuel will come from the USSR and will have a maximum enrichment of 20%, and so the reactors will have to be modified to cope.) A fourth research reactor, RA-4, has been imported from West Germany. In 1977 Argentina agreed to supply Peru with a 10 MW research reactor at a cost of $50 million. A zero-power reactor was supplied first and is now complete, and the 10 MW reactor is now under construction. The whole arrangement is covered by IAEA safeguards.

Since the mid 1950s Argentina has been producing uranium from her own reserves on a pilot scale. Fuel for the research reactors has been fabricated in Argentina, although the uranium has been enriched in the USA. In the late 1970s Argentina was producing a total of more than 100 tonnes of uranium per year from several uranium mills. Some of this uranium has been exported, notably to Israel. And in 1968 another nuclear fuel cycle facility was obtained, when a pilot-scale reprocessing plant began operation at Ezeiza. This safeguarded plant is capable of producing less than half a kilogram of plutonium per year. But in 1978 an announcement was made that work would begin soon on the construction of a commercial size reprocessing plant. (It is intended that plutonium recycling will be carried out.) This is also sited in Ezeiza nuclear centre near Buenos Aires. It is believed that the original reprocessing plant stopped operating in 1977.

Like India, Argentina planned to build a power research reactor (of about 40 MWe) in order to facilitate the scale-up from research reactors to commercial power reactors. This was cancelled, however, probably for reasons of economy.

Seventeen tenders were received from five nations for Argentina's first commercial nuclear power station, sited at Atucha, 110 km north-west of Buenos Aires on the Parana River. It was specified that tenders should allow a maximum amount of local participation in the project. The American bids, from General Electric and Westinghouse, were the lowest, but a natural uranium-fuelled heavy water reactor from Siemens was given preference, since it could be supplied from Argentina's own uranium reserves rather than relying on enriched uranium from the USA. The contract price for the Siemens reactor was DM280 million. Construction of the 320 MWe Atucha-1 reactor began in June 1968 and it became operational in 1974. It was essentially an extrapolation from a 52 MWe prototype operating at Karlsruhe. The heavy water necessary was supplied by the USA. The 50 tonnes of natural uranium fuel was prepared in Argentina, though the zircaloy cladding came from the USA. (In February 1980, it was announced that the USSR would build a plant to produce zircaloy cladding in Argentina.) The final cost of Atucha-1 was around $145 million and local participation amounted to about 40%.

Construction of a second nuclear power station began in 1973 at Embalse in Cordoba Province. (By this time the CNEA had grown into an organisation employing 3000 personnel, about 1000 of them being highly qualified. By 1978 the CNEA budget amounted to 243 billion pesos.) This Cordoba contract was won by AECL of Canada and Italimpianti of Italy, against competition from General Electric, Westinghouse and Kraftwerk Union (KWU). The Cordoba reactor is a 600 MWe CANDU (originally priced at $450 million) being built with local participation of the order of 50%. Canada's Export Development Corporation supplied a loan of $130 million to Argentina to finance the station. Cordoba is due to go operational in 1984.

Accusations were made in 1976 that payments of $1 million went to high ranking Argentine military officers to facilitate this deal. But the contract soon appeared unprofitable to the Canadians due to a large increase in the cost of heavy water production, and also the existence of a clause stipulating a maximum allowance for cost escalations. So, in 1977, the loss it was estimated AECL would suffer on the deal amounted to $130 million. Since the Argentine currency was also devalued, the whole contract was renegotiated. A cost-plus arrangement was substituted which increased the cost to Argentina by $100 million. (The total cost was put at $1.25 billion in August 1981.) The Argentinians still hoped, however, that a second 600 MWe CANDU could be built at Atucha (as Atucha-2). But the stringent safeguards policy applied by Canada from the mid 1970s onward was not acceptable to Argentina. (All of Argentina's reactors are, however, under IAEA safeguards, though the commercial reprocessing plant is not.) As a result, in May 1980, an agreement was reached between the CNEA and Germany's KWU over a 600 MWe heavy water reactor to be sited at Atucha despite the fact that Canada's tender was 50% lower than the German offer. It has been suggested that in order to land the contract German officials may have misled the Canadians into believing they too would demand full scope safeguards (Redick, 1981, p. 119). This Atucha-2 project is worth $1.5 billion and is due to come on line in 1987. Three more heavy water reactors and associated turbogenerators are likely to be supplied by KWU, to come on line in 1991, 1995 and 1997. These are likely to be sited in Mendoza province (western Argentina), Bahia Blanca, and on the Atlantic coast. The Swiss firm Sulzer is to supply a heavy water plant (at a cost of $300 million) with an annual output of 250 tonnes, in order to supply all these reactors. The heavy water plant is being sited at Arroyito in western Argentina and should begin production in 1984. At present Argentina has only a small heavy water plant, at Atucha, capable of producing 2–3 tonnes per year.

Argentina's uranium reserves are not known to any certain extent. But it is thought that there are reasonably assured, low cost resources of 20 000 tonnes and possible total resources of 40 000 tonnes. The agreement with West Germany over Atucha-2 also involves a deal on uranium exploration and mining through which 25% of any uranium co-operatively mined will go to West Germany.

Argentina has signed, but not ratified, the Treaty of Tlatelolco, a 1967 agreement to prohibit nuclear weapons in Latin America. The Treaty of Tlatelolco is currently in force for twenty-two Latin American nations, the most significant exceptions being Argentina, Brazil, Chile and Cuba. There are two additional protocols in the treaty, to be signed and ratified by some countries outside Latin America. Additional protocol one, whereby countries with territorial interests in Latin America pledge to keep their possessions free of nuclear weapons, has been agreed to by the UK and Holland but not by France and the USA. Additional protocol two, whereby nuclear weapons powers pledge not to use or threaten to use nuclear weapons against parties to the Treaty, has been agreed to by the UK, France, China, the USA and the USSR. (India was asked to ratify additional protocol two in 1975, but refused on the grounds that it had only detonated a PNE and therefore was not a nuclear weapons state.)

Argentina is not a signatory to the Non-Proliferation Treaty.

Brazil

Oil provides about 47% of Brazil's total energy supply, while biomass (30%) and hydro (20%) make up most of the remainder. Oil demand is currently around two million barrels per day. Oil production is fairly limited so that imports have been substantial. About 15% of oil consumed is produced locally, most of the rest coming from OPEC sources. Natural gas is imported from Bolivia. Proved reserves of oil amount to 900 million barrels and of natural gas to 1200 billion cubic feet. A substantial oil exploration effort is under way. There are oil shale reserves thought to be among the biggest in the world, containing the equivalent of about 800 billion barrels of oil.

In order to substitute alcohol for petrol in internal combustion engines, a government programme was instituted in 1974 to ferment sugarcane and to use the resulting alcohol in a 1 : 4 mix suitable for car engines with only minor modification. In 1979 alcohol accounted for about 14% of Brazil's car fuel consumption. It is planned that 1.6 million new cars will be built to run on alcohol alone in the 1980s.

As well as an abundance of hydroelectric potential, Brazil has large reserves of coal, at least 5 billion tonnes. But this coal is said to be of inferior quality, having an ash content as high as 55% (Miccolis, 1978), and it has therefore not been exploited to any large extent. Coal output in 1977 was around 10 million tonnes, but a plan for coal approved in 1979 intends that twenty-nine mines be opened by 1985 with a combined output of 35 million tonnes per year.

Most of Brazil's electricity-generating capacity (of around 27 000 MWe) consists of hydroelectric schemes. Oil-fired power stations provide 10% of capacity, coal 5% and hydro 85%. Brazil's total hydroelectric potential has been estimated at 209 000 MWe by the Ministry of Mines and Energy. It has been said by some observers that the majority of this potential lies in the Amazon basin in the north of the country, while most of the demand is in the south around Rio de Janeiro and Saõ Paulo. However, the Ministry's estimate is subdivided as follows: north, 96 000 MWe; north-east, 14 000 MWe; south-east, 55 100 MWe; south, 43 500 MWe. Of the potential in the south and south-east, 37% is either in operation or under construction. To provide electricity for the southern area of the country, Brazil has arranged to build a huge hydroelectric dam (Itaipu) jointly with Paraguay on the border between the two countries. This is to be the world's largest hydroelectric complex, having an output of 12 600 MW and costing $6 billion.

Brazil is currently in process of linking regional grid systems. The south and south-west grids are now interlinked, while transmission lines are under construction to link the north and north-west grids.

Brazil's interest in nuclear energy was evident in the 1950s when training of technical personnel began. An Atomic Energy Commission was hived off from the National Research Council in 1956. This Commission subsequently became known as the National Nuclear Energy Commission (CNEN).

During the early 1950s attempts were made to purchase a centrifuge uranium enrichment system from West Germany, and fuel processing plants from France. This was prevented by the USA, but suspicions had been aroused that Brazil was interested in weapons. In September 1957 a 5 MW 'swimming pool' reactor at the Atomic Energy Institute, Saõ Paulo, went critical. In 1960 the Institute of Radioactivity Research, Belo Horizonte, began operating a 100 kW research reactor, and in 1965 the Institute of Nuclear Engineering near Rio de Janeiro started up its 10 kW reactor, which had been constructed in Brazil. The enriched fuel for all three reactors was acquired from the USA, as well as the first two reactors themselves. The uranium itself originated from Brazil, however.

By 1967 the CNEN employed over 1100 people and was

spending between $6 million and $9 million annually. By 1979 the CNEN budget was about $33 million a year. In 1971, out of bids from Canada, West Germany, the UK and the USA, Brazil chose to purchase a 620 MWe reactor from Westinghouse. The site for this was on the coast at Angra dos Reis, 130 km south of Rio de Janeiro. The estimated cost of the project at the time was $266 million, of which just over half was to be supplied by a group of American banks led by the Export-Import Bank. The latter lent $69 million, and the other banks lent a total of $69 million together. Angra-1, as it is now called, has involved a fairly small amount of local participation (about 8%), consisting mainly of civil engineering work. It was about to begin operation in late 1981, fuel loading having been completed by September 1981. The uranium for this initial fuel loading was purchased from South Africa, converted to UF_6 by British Nuclear Fuels Ltd, and enriched in the USA at Oak Ridge. The used fuel is to be returned to the USA.

There was some delay in delivering the fuel for the initial core loading. Under the Carter Administration the US State Department and the Nuclear Regulatory Commission were reluctant to give a go-ahead after unsuccessful attempts to stop the transfer of reprocessing and enrichment technology from West Germany to Brazil (see subsequent paragraphs).

During 1974 the USA had reclassified previously firm contracts for enriched fuel supply as 'provisional'. This was mainly due to apprehension that there would be increased demand for nuclear fuel following the oil price rise, but no increase in enrichment capacity. This may well have been a contributory factor to the Brazilians' search for alternative sources of reactors and fuel, and indeed for their own enrichment plant (Perry and Kern, 1978).

A massive nuclear deal which may be worth $20 billion was made between Brazil and West Germany in 1975. (At the time it was estimated to be worth $5 billion and West Germany provided a loan of $1.7 billion; however, a 1981 official estimate of the cost between 1980 and 1995 quoted $18 billion.) This agreement was to involve the purchase, over the period up to 1990, of four pressurised water reactors, each of 1300 MWe, with an option on four more; a jet-nozzle enrichment plant with a capacity of 250–1000 tonnes SWU per year; a fuel fabrication plant having a capacity of 100 tonnes per year; and a reprocessing plant. (The jet-nozzle process is an unproven technology, thought to use about twice as much electricity as the gaseous diffusion method. It is specifically stated in the agreement that Germany gives no guarantee of success with this process. The more or less proven ultracentrifuge method could not be transferred to Brazil despite a 1969 agreement for joint co-operation on this technology because of objections from Holland and Britain, who jointly developed it with West Germany.) All these nuclear fuel cycle facilities, plus a conversion plant, are to be sited at Rezende, near Rio de Janeiro. Eventually the deal is to be paid off in part by Brazil supplying West Germany with enriched uranium.

This agreement was made contingent on safeguards being applied by the IAEA and on an assurance from Brazil that none of the nuclear equipment, materials *or know-how* would be used to make nuclear explosives. Additionally, replication of German-supplied facilities without safeguards, or their transfer to other nations, is prohibited. (At the present time all known nuclear facilities in Brazil are subject to safeguards, though not 'full scope safeguards'.) The full scale enrichment and reprocessing plants are only due to go operational when work on all eight reactors has been begun; that is, it is a condition of the agreement that Brazil purchase all eight reactors before the enrichment and reprocessing technology will be transferred. Meanwhile, pilot or demonstration plants will be built and the enriched uranium fuel for the German reactors will be supplied by URENCO – the British, Dutch and West German consortium. URENCO was to supply 200 tonnes of fuel per year over the period 1981–91, but this contract may be renegotiated to a lower supply figure. There are several reasons for this, including delays in reactor construction, a drop in the Brazilian nuclear budget (to $140 million in 1980, some resources having been diverted to hydroelectric development), and difficulty in supplying sufficient Brazilian natural uranium to URENCO. But, alternatively, it is also possible that URENCO may supply reload fuel for Angra-1 as well as the fuel for Angra-2 and Angra-3.

Construction of a pilot enrichment plant began in 1979, but was not scheduled for completion before late 1983. The full size, commercial scale plant would only be operational by 1987 or later. Plans for a pilot reprocessing plant were more or less complete in 1979 but construction work had not begun by late 1981. A laboratory-scale reprocessing plant was working, however, and Brazilians were undergoing training in reprocessing technology in West Germany. It appears that a commercial size facility will not be in existence until the very late 1980s at the earliest. As for radwaste disposal, no firm plans are yet in evidence.

Construction of the first two of these eight German reactors has already started. Angra-2 (1300 MWe) was begun in 1976 and is due in operation in 1987. Angra-3 (1300 MWe) was begun in 1978 and is due in operation in 1988. Neither of these was yet above ground level in 1981. The cost of these two reactors was originally thought to be at least DM3.7 billion, there being also between DM2 billion and DM3 billion allocated to associated equipment. A loan of $1 billion at concessionary interest rates has been organised by the Kreditanstalt für Wiederaufbau, a West German development bank. Lately, however, Furnas, the government-owned electricity board, has placed the cost of these reactors at $6 billion.

Brazil appears to be fairly committed to building two more of these 1300 MWe reactors, all four of which are to be constructed by Kraftwerk Union. The amount of local participation in these German reactors is to be around 40–50%, substantially higher than for the Westinghouse reactor. No sites have yet been selected for the third and fourth KWU reactors. Although all eight of these German reactors were to have been operable by 1995, it is now recognised officially that a target date of the year 2000 is more feasible. (In the early 1970s it was planned to have 60 plants producing 75 000 MWe by the year 2000.)

In 1978 a Brazilian parliamentary committee investigated alleged irregularities in the nuclear power programme, including questions of corruption and dubious site selection (Krugmann, 1981). The proceedings of this committee developed into a debate over whether the programme was sensible or not.

As an additional part of the Brazil–West Germany agreement, Urangesellschaft of West Germany is to help in exploration and exploitation of uranium, in return for which it is to have the option of buying 20% of any uranium exploited and a higher proportion in later years. Apparently, two sizeable uranium deposits had been discovered just before the agreement was made. Apart from the advantage of gaining a new source of uranium, West Germany would also manage, through the agreement, to circumvent the 'sensitive' issue of building a large uranium enrichment plant on West German soil. 'Sensitive' because it would be in close proximity to the Communist bloc and also in contravention of agreements made with the Allies after World War II.

Since 1969 Brazil has mounted a major effort in uranium exploration, and has expended close to $100 million for this purpose. As of December 1979 she was thought to have around 200 000 tonnes of reasonably assured and estimated additional resources. This is at least three times as much as India has, and the Brazilian grade is generally superior, being around 0.15% of U_3O_8.

Like India, Brazil is known to have large reserves of thorium. However, not a great deal of work has been done in Brazil regarding thorium utilisation. There was formerly a 'Thorium Group' consisting of about twenty scientists at the Institute of Radioactivity Research, Belo Horizonte. However, this was not an official research project, rather an expression of individual research interests. There was, though, a co-operative programme with France in this area, begun in 1967. Whether much information has been communicated from France to Brazil is not known.

Another agreement with France was announced in 1975. This covered the development of a fast breeder reactor to be built in Brazil and to be used for research purposes. The contract for construction of this reactor (called COBRA, an acronym for Co-operation Brazil) was thought to be worth $2.5 million. However, in 1981 equipment related to the use of sodium in fast breeders was being obtained from Italy at a cost of $11 million. (Brazil also had an agreement with Gulf/General Atomic Corp. for co-operation in R & D on a gas-cooled fast breeder to use thorium as a fuel.)

There is also collaboration with France in the building of a uranium mill and a uranium hexafluoride plant which are due to be completed in 1983 by the firm Pechiney Ugine Kuhlmann.

Brazil also has an agreement for nuclear co-operation with Iraq, signed in January 1980, under which the latter will get assistance in prospecting and producing uranium, and some reactor equipment and low enriched uranium. The whole agreement is to be subject to IAEA safeguards. (At the time Iraq supplied nearly half of Brazil's imported oil.)

In May 1980 a nuclear co-operation agreement was also signed with Argentina. Under this agreement some of the equipment for the German-supplied Atucha-2 reactor will be manufactured in Brazil by a joint Brazilian–German company. Information and training schemes will also be shared, and in later years Argentina will supply zircaloy to Brazil, and Brazil will supply enriched uranium to Argentina.

Brazil appears to have a fairly similar attitude to India regarding the subject of nuclear explosives and nuclear weapons. Government interest has long been expressed in the development of PNEs. In 1967 the CNEN commissioned a study on Brazil's ability to build a nuclear explosive. It was estimated that ten years or more would be needed to construct a bomb.

Brazil has signed and ratified the Treaty of Tlatelolco. (The Treaty of Tlatelolco does not prohibit PNEs.) But Brazil's full ratification has been made contingent upon the treaty's ratification by all Latin American nations (which has not yet occurred) and also contingent upon the signing of additional protocols to the treaty by states outside Latin America (which also has not yet occurred). Thus the Tlatelolco treaty is not in force as far as Brazil is concerned. The NPT is not in force for Brazil either. Brazil has long been an opponent of that treaty and has neither signed nor ratified it.

Brazil established a National Commission on Space Activities in 1961; this operates a research installation near Saõ Paulo and an active rocket-launching base south of Natal, near the easternmost tip of the country. Over 400 missile test launches were made between 1965 and 1974. There is thought to have been quite close co-operation between West Germany and Brazil on rocket systems. (The only other developing country to have its own launching facilities is India.)

Chile

Oil supplies 75% of Chile's commercial energy use and only about 20% of this oil is produced domestically. Recently, however, new oil deposits have been discovered in the Magellan Straits, so the government plans to increase production so as to meet 45% of oil requirements. It may even be possible to achieve self-reliance after the year 2000. Natural gas is also produced from the Magellan Straits and a liquified natural gas plant is in operation (this absorbed around 6000 million cubic metres in 1978).

Coal production in 1978 was around 1 million tonnes. The extent of reserves is rather uncertain, but a minimum of 4000 million tonnes has been quoted. Further discoveries are being made and the coal industry as a result is likely to go through a period of expansion.

Electricity generating capacity stands at 3000 MW, of which about half is hydro and half is thermal. Actual production of electric energy, however, amounts to two-thirds from hydro and one-third from thermal power stations. Total hydro potential has been estimated at 18 000 MW, and hydro capacity is likely to be steadily increased over the coming years as new projects come on line.

The Chilean Nuclear Energy Commission (CCEN) was founded in 1965. Two research reactors are currently operational. One is a British pool-type reactor of 5 MWth which began operating in 1974 at the University of Santiago. The other is a 20 MWth reactor, also near Santiago, built with help from Spain and completed in 1977. The enriched uranium for the first reactor is supplied by the USA, and for the second by France. In 1980 a national planning decision was made to go ahead with Chile's first nuclear power station (of 500 or 600 MWe), which is likely to be sited in the Santiago area. Substantial efforts are being made to locate uranium reserves. To date 5000 tonnes are thought to be available.

On the issue of proliferation treaties, Chile is not a party to either the Tlatelolco treaty or the NPT. The Tlatelolco

Conclusions 119

treaty has been signed but not fully ratified, and the NPT has
not been signed.

China

In 1977 the People's Republic of China derived its energy as
follows (Clarke, 1980, *passim*):

	Million tonnes coal equivalent
Coal	550
Oil	170
Natural gas	81
Hydroelectricity	9
	810

Approximately one-fifth of all the world's coal is now
produced in China. Estimated total reserves of coal are 1000
billion tonnes or more, enough to last for 2000 years at
present rates of use. Proved oil reserves are of the order of
20 billion barrels (enough to last thirty years), and proved
natural gas reserves are sufficient to last fifteen years. It is
likely that more reserves of oil and gas will be located in the
future.

All of China's oil is produced indigenously; in fact small
quantities of oil have been exported, principally to Japan.
Almost 55% of oil production comes from the Taching oil-
field in north-eastern China.

Given this abundance of fuels, China clearly does not
desperately need to embark on a massive nuclear power
programme. But one of the arguments in favour of going
nuclear is that most of the coal mines are situated far away
from demand centres: most of the coal is in the north, and
demand tends to be concentrated around coastal areas.
However, proven coal deposits exist in 1357 out of China's
2000 counties, but the quality and quantity of coal in
southern areas is low. Another argument for nuclear power
relates to pollution from coal-fired power stations; at present
no electrostatic precipitators are used, so such pollution is
considerable.

Electric generating capacity has been doubling about
every seven years of late. Total installed capacity in 1977
was about 40 000 MWe, of which around 15 000 MWe is
hydro and most of the rest coal-based (Clarke, 1980, p. 145).

The total hydroelectric potential could be 300–500 GW or even more. A lot of this potential is said to be in the west of the country, however, at a fairly considerable distance from the demand. China has five major regional grids which are not fully interlinked to form a national grid. The largest size of unit indigenously installed was 300 MWe in the mid 1970s and most high voltage transmission was at 220 kV. Industry consumes more than 80% of China's electricity.

In the early years after the revolution, China received a great deal of help from the USSR in nuclear developments. Since that time, emphasis has been placed mostly on nuclear weapons. The Russians helped to set up research reactors near Beijing and Lanzhou (around 1956), a gaseous diffusion enrichment plant at Lanzhou, and a uranium mine near Urumchi in Sinkiang. After the split with the USSR in 1958 China had to go it alone. It was in 1964 that China exploded a first A-bomb of 20 kilotonne yield and a H-bomb followed in 1967.

Currently there are several research institutes in China devoted to nuclear technology and atomic energy. These generally have basically simple research reactors such as swimming-pool reactors, though the atomic research centre run by the Chinese Academy of Sciences at Fangshan, 50 km south-west of Beijing, has a 10 MWth heavy water research reactor (originally supplied by Canada) in the process of being rebuilt at present, and a 3.5 MWth LWR. (This institute was set up in 1958, and now has about 1500 staff of whom half are scientifically or technically qualified.) Some of the equipment still shows signs of Russian influence in its design. The atomic energy research institute on the outskirts of Shanghai has, however, recently built a small reactor modelled after a Westinghouse core. (The Shanghai institute has 500 staff, about 60% of whom are scientists.) One of the latest research reactors to go into operation (at the end of 1980) is rated at 125 MWth and is located near the city of Chengdu.

As far as commercial nuclear power is concerned China is still at the planning stage. (China is the only nuclear weapons country without a nuclear power programme.) But in 1978 at the Fifth National People's Congress in Beijing, Mao's successor, Hua Kuo-feng, urged that nuclear power stations be established in China (Becker, 1978). Negotiations over two 900 MWe PWRs to be bought from France were cancelled by the Chinese in July 1979. These were to have been supplied by Framatome at a total cost of FF10 000 million. (American approval for this sale had had to be obtained, for the reactors were subject to Westinghouse licensing rights.) Apparently the Chinese withdrawal was necessary because they had concluded that they did not have the foreign exchange to pay for the project. (Though after the Three Mile Island incident in March 1979 the Chinese became very concerned about the safety of PWRs. They sent a team of scientists to the USA to investigate the accident, and this team became satisfied eventually that PWR safety was not an insuperable problem.) However, in 1981 this French contract was confirmed in principle (at a price of $2 billion), so it may be that the Chinese were merely doing some hard bargaining! The position regarding this project is still unclear because there is currently talk of producing a completely indigenous power reactor of around 250–350 MWe as an alternative to purchasing from abroad. The Shanghai atomic research institute has been working for some time on designs for a 300 MWe PWR, and work has also been going on on heavy water reactors and boiling water reactors. If the French contract does go ahead, the reactors may be sited in the Canton area so as to supply both Hong Kong and China with power. A feasibility study of this proposal has been completed jointly with the China Light and Power Company of Hong Kong, and the study is currently under consideration. Since 1978 Hong Kong and China have been co-operating over plans to interconnect their power systems (Leung, 1980). An agreement was signed in 1979 for Hong Kong to supply Shumchun (in China) with electricity. China will supply coal for the Castle Peak power station in Hong Kong which should go into operation in 1982. In return for power from a nuclear reactor, Hong Kong could provide harbour facilities for building and supplying the reactor. All this mutual co-operation no doubt anticipates the expiry (in 1997) of the lease held by the UK on the New Territories of Hong Kong, at which time they revert to China.

It is unlikely that China will have an operating power reactor before 1990; in fact a decision on whether to go nuclear and on what kind of reactor to go for might take several years to sort out. At the present the Chinese are trying to catch up to, or stay abreast of, modern developments in nuclear power technology.

It is believed that the Chinese have large deposits of uranium and thorium. One estimate puts uranium resources at 20 000–100 000 tonnes (World Energy Conference, 1974, p. 205).

China is not a signatory to the NPT.

Cuba

Cuba has been almost wholly dependent upon oil for its energy supplies, and 95% of its oil has been imported from the USSR. The oil sold by the USSR has generally been subsidized, i.e. below current world market prices, but in the late 1970s it was thought that Soviet prices would rise to become comparable to world prices. Offshore oil was discovered in Cuba in 1968 and annual output from this source is probably around 200 000 tonnes (very approximately). Installed electric generating capacity amounts to nearly 2000 MWe, almost all of this being oil-fired, the rest hydroelectric. Until 1973 there were two electrical grids, western and eastern, but in that year these were linked and a national grid formed.

A Cuban Nuclear Energy Commission (CEN) was formed in 1956 and in the late 1950s this organisation planned to obtain a research reactor from the USA and also a small (20 MWe) nuclear reactor to provide electricity for an agroindustrial complex. These plans had reached quite a firm, advanced stage when the Batista government was overthrown in a revolution and the projects were not carried out (Perez-Lopez, 1979). In the late 1960s, however, the USSR transferred a research reactor to Cuba, and in the 1970s an indigenous natural uranium, graphite-moderated research reactor was built.

A National Commission for the Peaceful Use of Atomic Energy (CEA) was set up in 1974, and in the same year an announcement was made in Moscow that design work had begun on a nuclear power reactor for Cuba. It was, though, only in 1979 that a firm announcement was made that a reactor would be built at Cienfuegos on Cuba's southern coast. This was to be a 440 MWe Soviet version of the PWR, using enriched uranium fuel from the USSR. All PWRs built by the USSR in eastern Europe and the Soviet Union have been without the containment building which is a standard feature of 'western' PWRs. They also appear to have either very limited or non-existent emergency core cooling systems. It seems likely that the Cuban reactor will be no exception in these regards. Work on this plant was due to begin in 1980, and it was thought possible that the USSR might aid Cuba in building six more of these reactors. Such an arrangement was conceived as usefully conserving the USSR's oil supplies, either re-allocating supplies for home consumption or freeing them for world trade and thus obtaining scarce foreign exchange. The agreement would not, in any real sense, lessen Cuba's dependence on her ally.

It is projected that a 300 MWe capacity pumped storage scheme will be built near to the reactor, using the Agabama river as the lower reservoir. Such a pumped storage scheme would assist in linking the power reactor to the variable demands of the grid.

Cuba has not signed either the NPT or the Treaty of Tlatelolco. Cuba refuses to sign the latter treaty while the USA maintains nuclear weapons in Latin America (e.g. at Guantanamo). It may be that delays in supplying the Soviet reactor have been due to attempts on the part of the USSR to persuade Cuba to agree to one or both of these treaties. The USSR is known to be in general favour of IAEA safeguards and it is highly likely that these will be applied to any equipment or material supplied to Cuba. In fact the president of the CEA announced in 1976 Cuba's intention to place all nuclear activities under IAEA safeguards, though by 1978 a safeguards agreement had not yet been signed.

Egypt
Major discoveries of oil were made in the mid 1970s, and by 1979 proven oil reserves stood at 300 million tonnes and production at 30 million tonnes per year. Natural gas has also been produced since the mid 1970s, and proven reserves are at least 75 million tonnes of oil equivalent. Egypt has some reserves of coal and could produce half a million tonnes a year by 1990.

Electric generating capacity is 4400 MW, about two-thirds being hydro and one-third being thermal. The Aswan High Dam makes up a very large part of the hydro capacity, its rating being 2100 MW. Total hydro potential is estimated to be only 4000 MW.

The Atomic Energy Organisation of Egypt was established in 1955. A research reactor of 2 MW was installed with aid from the USSR at Inchass (30 km from Cairo) in 1961. The USSR subsequently made more than one offer to sell a power reactor to Egypt and feasibility studies were carried out. In 1974 the USSR again made an offer, which coincided with an offer from the USA. The American offer was made to Egypt and to Israel, and involved the placing of safeguards on *all* nuclear facilities in the recipient country in return for the supply of 600 MWe LWRs. Egypt was prepared to accept this offer, but the Israelis declined. In 1976 the offer was formalised by the signing of a nuclear co-operation agreement between the USA and Egypt. Egypt hoped to obtain as many as eight reactors through this agreement by the year 2000. It appears though that the issue of proliferation in the politically sensitive Middle East was a stumbling block in the US–Egyptian negotiations. An application for a loan of between $600 million and $1 billion from the US Export–Import Bank was turned down. But Egypt ratified the NPT in March 1981 and this probably opened the way to complete agreement. It is possible that Egypt will now order reactors from France and from the USA. Negotiations have been in progress over two 900 MWe units from France, and a 600 MWe unit from Westinghouse. The likely site for the first nuclear reactor is Sidi Kreir, 30 km west of Alexandria.

Discovery of uranium was announced in 1977 and in 1979 the Nuclear Materials Corporation was planning to begin mining and processing uranium ore on an experimental, small scale basis. The ore was of a low grade and amounted to only 5000 tonnes.

Iran
Iran is one of the world's major suppliers of oil. Proven reserves in the mid 1970s amounted to 8200 million tonnes, or about 10% of world reserves. There are large quantities

of natural gas. Gas reserves in 1977 were estimated at 17×10^{12} cubic metres (about 14% of the world's total) and natural gas supplied 17% of Iran's energy consumption. In fact natural gas represents a greater energy source for Iran than oil, the former amounting to 100 billion barrels of oil equivalent and the latter 60 billion barrels (Kashfi, 1980). However, natural gas is less attractive than oil if they have to be transported, so the export market for gas is limited and less profitable. (There is, however, a 2000 km pipeline supplying 10^{12} cubic metres of gas annually to the USSR – this is about equal to indigenous consumption.) As a result, in recent years, about 45% of all gas extracted in Iran has been flared. Such gas might, less wastefully, have been used for power generation.

There are also substantial reserves of coal in Iran, estimated at more than 1000 million tonnes. Clearly, Iran does not need nuclear power to alleviate an energy shortage. Electric generation capacity stood at 5300 MW in 1978. Of this, about 1000 MW was hydro and total hydro potential has been put at 10 000 MW.

Iran's first involvement with nuclear research came with the establishment of a Nuclear Centre at Tehran University in 1959. This was founded with American co-operation. An American made, IAEA-safeguarded, 5 MW swimming-pool type reactor was built there, and this became operational in 1967. It was in 1973 that the Atomic Energy Organisation of Iran was set up, replacing the earlier Atomic Energy Commission.

By the mid-1970s the Iranian AEO had a staff of 150 scientifically qualified personnel, 90% of them foreign. Half of these foreign personnel were from Argentina, the reason being primarily that Admiral Oscar Armando Quihillalt, who was for fifteen years head of Argentina's nuclear programme, became, through the auspices of the IAEA, chief consultant to the Iranian AEO.

In the mid-1970s orders were placed for power reactors, all of them PWRs. The rationale for developing nuclear power was that oil prices would continue to rise while the cost of nuclear power would remain stable. Thus, it was argued, it would be better to export oil than to use it domestically, and use the resulting revenue to finance a nuclear power programme. (The weapons issue does not seem to have been a factor.)

Two reactors of 900 MWe each were ordered from the French company, Framatome, to be built in south-western Iran. Construction began in 1977 at Darkhovin on the Karun River, between Ahwaz and Abadan. The combined cost of these reactors was reckoned to be $3 billion at the time the contract was signed. And two reactors of 1200 MWe each were ordered from KWU to be built at Bushehr on the Persian Gulf. Construction of these KWU plants began in 1976. The KWU order was said to be worth $4 billion at the time. Both orders included fuel supplies for the first ten years of reactor life. 'Commissions' on these purchases were alleged to have amounted to 20% of the total contracts, i.e. some several hundred million dollars per reactor (Mossavar-Rahmani, 1980).

To ensure a supply of enriched uranium for these reactors, Iran lent the French Atomic Energy Commission $1 billion over a fifteen year period in exchange for a 10% share in the EURODIF gaseous diffusion enrichment plant which was being built at Tricastin. Other shareholders in this scheme, apart from France and Iran, are Belgium, Spain and Italy. This plant began operation in 1980. Iran also had a 25% share in the COREDIF gaseous diffusion enrichment plant, again along with France, Belgium, Spain and Italy, but the future of this plant is, at present, doubtful, partly because Iran has reconsidered its commitment to the project.

Attempts were made to reach agreement with the USA also, over the purchase of reactors and supplies of fuel. Provisional agreements were made to purchase between six and eight reactors and enough fuel to supply them, but satisfactory safeguards clauses were never worked out. In 1977 the Shah of Iran was said to be ready to sign a $6 billion contract for four more French reactors. And in the same year a letter of intent was issued for four more plants from KWU worth $11 billion in total and to be sited 170 km south of Tehran.

These tentative orders were designed to fulfil a goal of 23 000 MWe of nuclear power by 1994 (which had been set in 1974). This figure had been arrived at by assuming that in twenty years' time electricity consumption would be at western European levels, and that 40% of total capacity would be nuclear. The cost of this 23 000 MWe programme was put at $24 billion in a study by the Montreal Engineering Company, which reported to the Iranian government in 1975. But by 1978 the estimated cost of such a programme had risen to between $80 billion and $120 billion.

In particular the estimated cost of the two German reactors under construction had risen to $7.5 billion total, and of the two French reactors to $6.4 billion, i.e. over $3000 per kilowatt of electricity in both cases.

Since the revolution which deposed the shah, all these plants have been cancelled, including those which were under construction. Iran cancelled the two French plants in April 1979, at which stage construction had not progressed beyond site work. In 1981 Framatome was suing Iran for alleged withholding of payment for work completed. After Iran indicated that the German contract might also be cancelled, KWU itself terminated their contract in July 1979, citing Iran's failure to pay $450 million for work done to date as a reason. (Iran had already paid $2.75 billion towards the Bushehr plants at that time.) One of the German reactors was thought to be about 80% complete as far as civil engineering work was concerned, though the reactor core itself was really only just begun. The cost of that plant was said to have already amounted to $3 billion. The other German plant was around 50% complete in civil engineering terms. In 1982 it seemed the Khomeini regime intended to renew the French and German contracts, though with enhanced Iranian involvement.

Iran mounted a major uranium exploration effort until the revolution intervened. Important deposits were rumoured to have been found in 1976, though the fact that Iran purchased 28 000 tonnes of uranium ore from a number of different countries would seem to indicate a different story. South Africa supplied 14 000 tonnes of uranium oxide at a price of $700 million, in return for Iran putting up part of the money for the building of a uranium enrichment plant in South Africa.

Iran is a party to the NPT, having signed and ratified it.

Iraq
Iraq is, of course, one of the world's major exporters of oil. Commercial energy supplies in Iraq in 1978 were made up

of oil, 73%, and natural gas, 27%. About 85% of natural gas produced was flared in the late 1970s.

As of 1978 total electrical capacity amounted to only 860 MW, all of this being oil- or gas-fired. However, over 1000 MWe of hydroelectric capacity and 3000 MWe of thermal capacity were under construction during the early 1980s (Lloyds Bank, 1981).

France agreed in 1976 to supply Iraq with two research reactors to be housed in the same complex 15 miles from Baghdad: one an Osiris-type of 40 MW with a maximum rated capacity of 70 MW (renamed Osirak) and the other an Isis-type of 500 kW. Part of the agreement included the supply of 72 kg of 93% enriched uranium fuel – though at any one time only sufficient fuel (about 20 kg) to load the reactor cores would actually be available to Iraq. The spent fuel was to be returned to France for reprocessing. The value of this reactor deal has been variously estimated at $300 million. The delivery of the Osirak reactor was delayed after the construction plant at Seyne-sur-Mer was sabotaged with explosives in 1979. This sabotage, and the murder in Paris of an Egyptian scientist, Professor Yahia El Meshad, who headed the Iraqi nuclear programme, have, speculatively, been put down as the work of Israeli security forces.

Suspicions that the Iraqis were strongly interested in nuclear weapons manufacture (despite the fact that they have ratified the NPT) were aroused on two counts. First, they contracted to buy a reprocessing plant, on a laboratory scale, from an Italian company based in Rome. Second, they imported supplies of UO_2 from Brazil, and 140 tonnes and 200 tonnes respectively of yellowcake from Portugal and Libya. It was speculated that Iraq's intention was to irradiate this uranium around the Osirak core, then extract the plutonium so produced. In this manner it might have been possible to make enough plutonium for about one bomb per year. The IAEA asked to institute more frequent inspections than it had up to then been carrying out (every fortnight rather than every three months). In consequence of these suspicions, the Israelis carried out a pre-emptive raid in June 1981, bombing the reactor from F-16 aircraft and effectively destroying it, before it had become operational. Negotiations have been in progress between Iraq and France over a replacement reactor (Saudi Arabia having offered to provide $300 million finance for it.)

At one time it also appeared likely that Iraq would buy a 900 MWe PWR from France, but apparently negotiations were broken off due to the exigencies of war with Iran.

Israel

Israel imports virtually all of its commercial energy in the form of oil (plus very minor quantities of coal). Indigenous supplies of oil and natural gas have been found to be extremely limited. Solar energy provides 1.1% of total energy supply.

Electricity is virtually all generated in thermal power stations. Capacity amounted to 2500 MWe in 1978.

An Atomic Energy Commission was established in 1952 under the Ministry of Defence, and this was later reorganised and placed under the direct control of the Prime Minister in 1966. Nuclear co-operation with France began in 1953, and during the late 1950s and early 1960s a 26 MW natural uranium research reactor was built at Dimona with French assistance. This reactor is not subject to safeguards. Collaboration with France terminated in 1969.

Technical co-operation with the USA began in 1955 and as a result a 5 MW reactor was built at Nahal Soreq, south of Tel Aviv. The highly enriched fuel for this reactor was also supplied by the USA. The reactor and fuel are covered by IAEA safeguards.

The Israeli government approved the completion of a power reactor (a 900 MWe LWR) by 1984, but this project has now been postponed indefinitely. It seems that uranium supplies are regarded as a problem, since Israel has discovered none itself despite searching since 1950. Canada and Australia would be unlikely to supply a country that has, it is thought, a strong interest in nuclear weapons, but the USA and South Africa might oblige. Planning studies on the power project are continuing nevertheless.

A reprocessing plant capable of separating out weapons-grade plutonium from spent fuel is in operation at Dimona. By the mid 1970s it was believed that Israel was likely to have an arsenal of between ten and twenty nuclear weapons

– plus the planes and missiles to deliver them (Lefever, 1979, p. 64). Israel has not signed the NPT.

Mexico

Commercial energy in Mexico is supplied from the following sources: oil and gas, 88%; hydro, 7%; coal, 5% (Eibenschutz *et al.*, 1977). Mexico is now estimated to have the sixth largest proven reserves of oil in the world (40 000 million barrels) and is likely to have total resources of around 200 000 million barrels. At the end of 1979, oil production stood at 2 million barrels per day. Natural gas is also available indigenously, proven reserves being put at 30 000 billion cubic feet. Proven coal reserves amount to 675 million tonnes, with total reserves probably coming to 12 000 million tonnes. It is planned that coal-fired power stations will be set up during the 1980s. At present, of the country's total electrical capacity of 12 000 MW, just over half is oil- and gas-fired, while just under half is hydro-electric. Total hydro potential has been estimated at 20 000 MW.

The National Commission on Nuclear Energy was established in 1955. Two research reactors, each of 1 MW, were set up in the 1960s. Fuel for these research reactors, and quite a lot of technical assistance, came from the IAEA.

In the late 1960s plans were formulated for a nuclear power plant to serve Mexico City, to be sited at Laguna Verde in the state of Veracruz. Bids for the construction of this plant were received from the USA, the UK, Canada, the USSR, Japan, Sweden, West Germany, France and Italy. Seven companies from the USA, the UK, Canada, Japan and West Germany were selected to take part in a final round of bidding. In September 1972 it was announced that General Electric had been awarded the contract for a 660 MWe reactor, with financial backing coming from the US Export-Import Bank. Work began at the Laguna Verde site in 1972, but was suspended for four years due to changes in government policy. Now, two identical reactors are being built at this site, and it is hoped that the first of them will be finished in 1984. The latest estimated cost of these two reactors is $1.91 billion. Eventually, a third and similar unit is scheduled to be built at the same site.

A tender for a further 2400 MWe of nuclear power was put out by Mexico in April 1981, and the intention was announced of going for 20 000 MWe of capacity by the year 2000. A request for technology transfer covering the entire fuel cycle (including enrichment) was made a part of the bidding instructions for reactor tenders.

A pilot reprocessing plant was under construction in the late 1970s, with, it is said, IAEA assistance. Studies have been carried out on the technical and economic feasibility of establishing a commercial size reprocessing facility. This would be subject to IAEA safeguards according to Mexican government officials. There is also a small fuel fabrication plant and plans for a larger version in future years.

Mexico has proven uranium reserves of 11 000 tonnes and has possible additional reserves of 300 000–500 000 tonnes. However, for the Laguna Verde plant, uranium is being purchased from France and enriched in the USA, because Mexico is without the necessary technology to achieve this. (Mexico originally applied to the IAEA for enriched uranium, and this was the first time that body had had a request for fuel for a power reactor. Through the IAEA the current arrangement was worked out.) It is said that the uranium for further power reactors will be domestically produced. In May 1980 a co-operative agreement was signed with France covering nuclear research and prospecting and exploitation of uranium.

There appears to be little official interest in Mexico in nuclear weapons, and that country is a party to the NPT and the Treaty of Tlatelolco.

Pakistan

Coal, oil, natural gas and hydroelectric potential are all indigenous to Pakistan. Coal production has amounted to around 1.5 million tonnes per year and imports have been necessary to supplement this. Coal reserves have been estimated as around 260 million tonnes. Oil was discovered in the mid 1970s and domestic production soon made up about 15% of total oil supply. Natural gas reserves have been estimated as 23 million million cubic feet, while production in 1978–9 was 214 000 million cubic feet.

In 1976 Pakistan's energy supply was made up as follows:

oil, 40%; natural gas, 35%; hydro, 17%; coal, 6.4%; nuclear, 1.7%; liquified petroleum gas, 0.3%. Electric generating capacity in 1979 amounted to more than 3000 MW, of which about half was hydro and half was thermal. A major expansion of hydroelectric capacity has been intended with the commissioning of the Tarbela Dam, which will develop 2000 MW when complete. Total hydro potential was estimated in the 1960s to be 20 000 MW.

As early as the mid 1950s, when the Pakistan Atomic Energy Commission was set up, a nuclear training programme was begun under which the USA, the USSR, the UK, France, West Germany, Canada and the IAEA provided training for Pakistani personnel. By the early 1970s the number of qualified nuclear scientists and engineers in Pakistan was nearly 600.

A research reactor was purchased from the USA in the early 1960s. This was a 5 MWth swimming-pool type, built at the Pakistan Institute of Nuclear Science and Technology, near Rawalpindi, for which the USA also supplied the 90% enriched fuel. During the year in which this reactor went critical, 1965, an arrangement was made with Canada to purchase a 125 MWe heavy water power reactor to be sited near Karachi. (The agreement between Canada and India over the purchase of RAPS-1 had been firmly fixed by this time.) Construction of this reactor began in 1966 and it went into commercial operation in 1972. The Karachi nuclear power plant (KANUPP) was a turnkey project with Canadian General Electric as the prime contractor and Montreal Engineering as subcontractor for the 'conventional' side of the plant. The fixed price contract for the complete plant was about $60 million. In general outline the plant was similar to that at Douglas Point, and was essentially designed for base-load operation. However, the plant control provided for automatic load following because the Karachi Electric Supply Corporation could not use the plant's full output at night and over weekends. Both of these reactors (the research reactor and the power reactor) are subject to IAEA safeguards. IAEA inspectors were visiting the KANUPP plant on a monthly basis in 1981.

In December 1976, the Canadians cut off all nuclear co-operation with Pakistan, so that when KANUPP broke down the Chinese provided technical assistance in August 1977. Supplies of fuel for KANUPP from Canada were also cut off at the end of 1976. So Pakistan set up its own fuel fabrication plant at Chashma to manufacture fuel from domestic uranium supplies. This now provides sufficient fuel to keep KANUPP in operation. It appears though, that Pakistan has very little uranium of its own since uranium has been purchased directly and indirectly from Niger. Niger takes a share of the output from the French-run mines within its borders and sells uranium not only to Pakistan but to Libya, which in turn sells to Pakistan. Neither Niger nor Pakistan has signed the NPT, but Pakistan has made a statement that the uranium would not be used for the manufacture of an explosive device (Libya *has* ratified the NPT).

A second power reactor is scheduled to be built at Chashma; this would be a 600 MWe LWR to come on line by the end of the decade. In 1981 it appeared that Spain would probably make a tender for this unit.

Regarding fuel cycle services: the heavy water supplies for KANUPP have come from the USA, but Pakistan has begun work on its own heavy water plant to produce 13 tonnes per year. When, in 1976, France agreed to sell Pakistan a reprocessing plant at a cost of $150 million (Tahir-Kheli, 1978) the USA and Canada made disgruntled noises. Not long after, in December 1976, Canada terminated all nuclear exports and technical aid to Pakistan, citing Pakistan's failure to sign the NPT and the proposed reprocessing plant as particular reasons for this action. At the time Pakistan protested that she was interested in a reprocessing plant purely for reasons of energy supply (to recycle fissile fuel and, in the future, to supply fast breeder reactors). Pakistan agreed to allow inspections by French and IAEA safeguards officials and to have similar supervision of any other plant built by Pakistan and using the French reprocessing method. However, after France withdrew from the arrangement in 1978 (probably partly due to American pressure and partly due to the relative smallness of the plant and the contract price) Z. A. Bhutto, the former Prime Minister, admitted that the plant was desired for weapons purposes. In fact, much earlier, in 1966, Bhutto himself had said that if India acquired a nuclear weapon, Pakistan would also do so, even if the population had to 'eat grass' (Khalil-zad, 1980, p. 12). A financial penalty to compensate for the cancellation should have been due, but perhaps this was waived since France had already delivered blueprints for the plant. Pakistan attempted after 1978 to obtain components for a reprocessing plant from an Italian subsidiary (ALCOM) of one of the French firms formerly involved, until this was stopped by France.

Having failed to obtain a reprocessing plant, Pakistan set out to get an enrichment plant instead. Kahuta, near Rawalpindi, was thought to be the proposed site, but now a conversion plant for production of UF_6 (bought from West Germany) and an enrichment plant are thought to be sited underground at Multan, 320 km south of Rawalpindi. Apart from strategic reasons for a move, it seems that Multan will provide less in the way of electricity transmission problems, there being a 260 MWe power plant nearby. (A moderate size enrichment plant might easily absorb 200 MWe.) It is suspected that Libya has provided some financial backing for the enrichment project, amounting to more than $100 million. Libya at one time attempted to purchase a bomb from France, China and India but was rejected by all of them. Under the present arrangement with Pakistan, Libya probably expects to receive a bomb in return for its financial support. During 1975–7 a number of 'front organisations' had been established for the purpose of buying crucial components in the UK, Holland, Switzerland, West Germany and the USA. Whether enough items were procured before this strategy was uncovered is not clear. However, it seems feasible that enough centrifuges could be set up to make enriched uranium for about six weapons per year. The Pakistanis do have quite a lot of technical know-how in this subject, for one of the leading lights in the project is Abdul Qadir Khan, a scientist who worked at the enrichment plant at Almelo, Holland, before returning to Pakistan. Only the next few years will prove whether the Pakistanis can be successful in this venture or not. The USA terminated all non-food aid in 1979 because of the proposal to build an enrichment plant, but after the revolution in Iran and the invasion of Afghanistan, military assistance was renewed. It is suspected that bomb attacks against foreign collaborators with the Pakistani programme are the work of Israeli agents – these operations could act as a brake on the project.

It seems likely that almost all of Pakistan's ventures in the nuclear field have been motivated by a desire to keep up with India. This view would seem to be confirmed by a statement made in recent years by General Zia ul-Haq, to the effect that southern Asia should be made a nuclear-free zone and all nuclear facilities in India and Pakistan opened to international inspection. A similar statement was made by Bhutto as early as 1972 (Kapur, 1980). These proposals have been rejected by India.

However, an alternative motivation for Pakistan could be that of using the mere threat of a nuclear weapons programme as a bargaining device in order to obtain more supplies of conventional arms from her ally, the USA. Restrictions on the nuclear programme would be 'exchanged' for conventional arms and the hoped-for goal would be closer parity to India in terms of military power. This would be a tricky game to play, though, and is just as likely to diminish arms supplies as to increase them.

Philippines

The Philippines has traditionally obtained 95% of its energy from imported oil. However, commercial oil discoveries were made offshore near Palawan Island during 1977–8. Production of 40 000 barrels per day began in 1979, and it is hoped that self-sufficiency can be achieved by the late 1980s.

The rise in the price of oil during 1973 led to an emphasis upon developing more indigenous energy supplies, particularly hydroelectric and geothermal potential. Hydroelectric capacity is projected to rise from 1000 MW to 3800 MW by 1988. (Total hydro potential has been put at 7500 MW.) Twenty-four new geothermal power stations have been planned with a capacity in all of 220 MWe. These are to be in operation by 1985.

The exploitation of coal was thought to be unwarranted until recently. A new survey has estimated reserves at 2000 million tonnes. Much of this is low grade coal, but it is possible to mix this with imported Australian high grade coal and to burn the mixture in power stations. Production of coal amounted to only a quarter of a million tonnes in 1978 but is expanding rapidly. Output in 1988 is projected to be over 4 million tonnes.

Total electric generating capacity stands at over 4000 MW. This is made up as follows: oil-fired stations, 70%; hydro-electric, 25%; geothermal, 5%.

The Philippine Atomic Energy Commission was established in 1958. Subsequently a 1 MW swimming-pool type research reactor was installed in 1963. During the late 1960s and early 1970s studies were made on the feasibility of setting up a power reactor. In February 1976 a contract was signed with Westinghouse for a 620 MWe PWR with turbo-generators. This reactor is being built on the Bataan Peninsula 70 km west of Manila. It is a replica of a plant being built in the USA. It is now due to go on line in late 1985. The final cost of this reactor was put at $1.1 billion in 1978 but now it may rise to $1.9 billion. Foreign loans of $800 million have been made available to finance it. The availability of a $664 million US Export-Import Bank loan was put in doubt for a time due to allegations that an illegal commission of up to $50 million had been paid to Herminio Disini (a friend of President Ferdinand Marcos) who arranged the deal (*Far Eastern Economic Review*, 23 June 1978, p. 97).

Another disruption to this project occurred after the Three Mile Island incident at Harrisburg in the USA in early 1979. The Philippine government became unhappy about the safety of their PWR, despite the fact that the Three Mile Island reactor was designed and built by Babcock and Wilcox rather than Westinghouse. The government felt it did not receive sufficient assurance from Westinghouse that the Bataan reactor would be safe, and so it halted construction work. It was only in September 1980 that President Marcos allowed construction to continue after a fifteen month delay, Westinghouse having agreed to instal various safety features such as an earthquake monitoring system and filter systems to catch volcanic ash (there is a volcano quite near to the reactor site, which is located in a seismic area).

Known reserves of uranium in the Philippines are rather small (less than 1000 tonnes) and the fuel for the Bataan reactor will be imported. Australia has agreed to supply uranium for Philippine use after a 'safeguards' arrangement was worked out in 1978. This uranium will presumably be enriched and fabricated into fuel in the USA. Exploration for local uranium deposits is continuing.

The Philippines is a party to the NPT, having signed and ratified that treaty.

South Africa

South Africa has considerable coal reserves (44 000 million tonnes) but no oil. Due to potential difficulties over obtaining oil from most of the supplier nations, South Africa extracts oil from coal and is the only country in the world to do this on a commercial scale. At present about 10% of South Africa's petroleum fuel is supplied by this method, and when further plants open in the mid 1980s this should rise to about 50%. The plant at Sasolburg produces mainly petrol and diesel fuel, and absorbs about 1.6 million tonnes of coal per year. (Total annual coal production amounts to more than 100 million tonnes.)

Electric generating capacity stands at 17 600 MW, of which 17 000 MW is thermal (almost all coal-fired) and 600 MW is hydro. Total hydro potential has been estimated at 5000 MW (this is a surprisingly low estimate).

The South African Atomic Energy Institute was founded in 1949 and later replaced by the Atomic Energy Board (AEB) in 1957. South Africa was one of the founder members of the IAEA in the mid 1950s. During the 1950s twenty-seven uranium mines and seventeen uranium mills were opened. All the mines extracted uranium as an adjunct to gold mining. Most of the uranium was destined for the USA and the UK, but some went to France, West Germany, Japan and Switzerland. Currently South Africa is reckoned to have at least 300 000 tonnes of low cost uranium or about one-fifth of the world's reserves.

During the late 1950s and early 1960s nearly a hundred South African scientists and engineers received training in nuclear technology in the USA.

There are two research reactors in South Africa. The first, Safari 1, went critical in 1965. It was designed and constructed by the Allis Chalmers Corp. of the USA. It uses 90% enriched uranium (supplied by the USA) and has a capacity of 20 MWth. The second research reactor, Safari 2, went critical in 1967. This reactor is of South African design and construction, but the heavy water and 2% enriched uranium fuel come from the USA. Both reactors are at the National Nuclear Research Centre, situated at Pelindaba near Pretoria. The majority of the AEB's 1900 employees work at this centre.

In 1971 the Electricity Supply Commission of South Africa (ESCOM) wrote to seventeen organisations advising of its intention to ask for tenders for a 500–600 MWe reactor. Seven consortia were placed on an approved list for a 1000 MWe reactor in 1974, and this was narrowed down to a short list of three in 1975. The consortium of Getsco–Brown Boveri Construction–Benucon was selected in 1976 as the preferred tenderer (with fuel to come from GE). However, this was subject to foreign government permits regarding export licences and export credit. Assurances over such permits being made available were received from the Swiss and US governments, but the Dutch government did not confirm any risk coverage for the substantial export financing involved. As a result, ESCOM turned to the second choice on its short list, a French consortium.

So, in 1976 construction work began on two PWRs, named Koeberg power station, at Duynefontein on the South Atlantic coast, 27 km north of Cape Town. These reactors each have a capacity of 922 MWe and the contract for them was worth $1000 million in 1976. Framatome is responsible for the reactors, steam generators and fuel-handling facilities, Alsthom–Atlantique for the turbo-generators and electrical equipment, and Spie–Batignolles for the civil engineering. Some of the conventional equipment for the station is being manufactured in South Africa itself. The plant is being built on special earthquake-resistant foundations. Eighty-two per cent of the finance for Koeberg 1 and 2 has been supplied by a group of French banks headed by the state-owned Credit Lyonnais. It is hoped the reactors will go into operation in 1982 and 1983 respectively. The fuel was to be supplied by the USA and would consist of 72 tonnes of 3% enriched uranium in each core. The reactors and the fuel are covered by trilateral safeguards agreements (between South Africa, France and the IAEA over the reactors, and between South Africa, the USA and the IAEA over the fuel). It is stipulated that spent fuel must be reprocessed in the USA. However, throughout the late 1970s and early 1980s the USA held actual supply of enriched fuel in abeyance while attempts were being made to persuade South Africa to sign the NPT and accept IAEA safeguards on its pilot enrichment plant. These attempts had not been successful by November 1981. In the event of complete breakdown of talks, it seemed possible that South Africa would either attempt to obtain fuel from France (in which case similar pressure would probably be applied to sign the NPT and/or accept safeguards) or would hold back Koeberg's operation until indigenous fuel was available. However, in August 1981 South Africa decided to deliver uranium feed to the Portsmouth, Ohio, enrichment plant, despite having no assurance of an export licence for enriched uranium.

As mentioned before, South Africa has several facilities for producing U_3O_8. In the mid 1970s a plant for conversion to uranium hexafluoride was opened and, at the same time, a small uranium enrichment plant began operating at Valindaba; this cost $115 million to set up (Adelman and Knight, 1979). This enrichment plant uses a modified version of the German jet-nozzle process and is capable of producing 50 tonnes per year of 3% enriched uranium, at present, but is to be expanded to a capacity of 200 or 300 tonnes per year by the mid 1980s. There have been several rumours to the effect that South Africa has a small reprocessing plant, but no definite proof.

In 1977 American and Soviet satellites detected structures in the Kalahari Desert which were reckoned almost certainly to be preparations for a nuclear explosive test by South Africa. Apparently, however, this site was not in fact used for a test. (Andrew Young, the US Ambassador to the UN, ruled out a ban on supplies of nuclear fuel to South Africa, saying that in order to monitor the situation it was necessary 'to keep some kind of relationship'.) On 22 September 1979 an American satellite detected a double flash of light in the South Atlantic near South Africa, and this flash was interpreted to be characteristic of an atmospheric nuclear explosion of under 4 kilotonnes. South Africa vigorously denied that she had carried out a nuclear test, but a convincing alternative explanation for the occurrence has not been found. The American satellite was specifically designed to detect nuclear explosions and had monitored forty-one flashes over the previous fifteen years, all of which were confirmed to be nuclear tests in the atmosphere conducted by the French and the Chinese (Smith, 1980).

The 1977 and 1979 incidents in combination make it highly likely that South Africa does now have a nuclear weapon. South Africa does not accept the NPT.

South Korea

Total energy use in South Korea amounts to about 50 million tonnes of petroleum equivalent per year. This is made up as follows: oil, 62%; coal, 25%; wood, 8%; the rest is hydro and nuclear. All the oil is imported, but there are hopes that oil will be discovered offshore. Coal reserves are at least 1300 million tonnes, some of it of good quality. Production of anthracite in 1978 amounted to 18 million tonnes. Some coal is, nevertheless, imported. Total electric generating capacity is more than 6000 MW, of which the majority is thermal. Hydro potential has been put at 5500 MW, against the current capacity of 700 MW.

South Korea's interest in nuclear energy dates back to 1956 when a co-operation agreement was signed with the USA. In 1957 Korea became affiliated to the newly formed IAEA and in 1959 an Office of Atomic Energy and an Atomic Energy Research Institute (at Seoul) were set up. During the 1960s the latter was involved mainly in basic research, but in the 1970s it ran a manpower training programme and worked on developments in nuclear power technology. These days the Atomic Energy Research Institute has 700 staff, and the Korean Nuclear Fuel Development Institute has 400.

Recently the nuclear industry in South Korea was re-arranged along roughly similar lines to that in the USA. An Atomic Energy Commission has executive responsibility, while a Nuclear Regulatory Bureau applies safety standards, etc.

A first research reactor was installed in 1962; this was a 250 kW Triga Mark 2 reactor. A 2 MW Triga Mark 3 reactor went critical in 1972. Meanwhile, in the late 1960s, a decision was made to opt for nuclear power, and a 587 MWe reactor was ordered from Westinghouse, to be sited at Pusan on the south-eastern coast. This reactor, Kori-1, was commissioned in November 1977 and went into full commercial operation in March 1978. (English Electric supplied the turbogenerators, somewhat behind schedule.) The cost of this reactor was put at $320 million and about 20% of the equipment and materials for its construction were supplied within South Korea (Cha, 1978). (It is planned that local supply for the four subsequent reactors will grow to 30–40%.)

Further reactors were ordered during the 1970s, as shown in Table 6.5.

The reactor at Wolsung is essentially a replica of Canada's Gentilly-2 (Lee *et al.*, 1977), and the contract for it includes fuel from Canada. This reactor came with a Canadian Export Development Corporation loan of $330 million. (It was estimated in 1975 that its final cost would be $700 million and that Kori-2 would be in the same price range.) US Export-Import Bank loans have been agreed for the other reactors, as follows: Kori-1, $65 million; Kori-2, $285 million; Kori-3, $398 million; Kori-4, $431 million; Gyaema-ri 1 and 2, $936 million.

In early 1981, after five years of negotiations, an agreement was finalised for the sale of two 900 MWe PWRs from Framatome of France, apparently at a bargain price of $1 billion, including ten years' fuel supply. Bids for the supply of the turbogenerators for these reactors were made by

Table 6.5 *South Korean reactors*

Name	Supplier	Size (MWe gross)	Construction started	Operational
Kori-2	Westinghouse	650	1977	end 1983
Wolsung-1	AECL	679	1977	1982
Kori-3	Westinghouse	990	1978	1984
Kori-4	Westinghouse	990	1978	1985
Gyaema-ri-1	Westinghouse	997	1979	1986
Gyaema-ri-2	Westinghouse	997	1979	1987

Wolsung is being built at Naah-ri near Ulsan, 50 km north of Pusan, and Gyaema-ri is near Kwangju on the south-western coast.

142   *Nuclear Power in India*

Westinghouse, General Electric, Alsthom Atlantique, Brown Boveri of Switzerland, Mitsubishi, and General Electric Company of the UK.

The Koreans were particularly keen to sign a contract with France, for this has several advantages for Korea. First, they diversify away from dependence on the USA (for enriched fuel and reactors). Second, they maintain links through which they might eventually obtain a reprocessing plant (see the later part of this section for an account of Korea's interest). Third, they may be able to obtain fast breeder technology through the French. Reprocessing and fast breeders could reduce the need to import uranium and give a greater degree of independence (Parrott, 1980).

France had been due to supply a reprocessing plant (negotiations over this had been carried out in the mid 1970s), but this was cancelled, mainly due to American pressure after concerns over proliferation. France and South Korea did claim that it was their own decision to cancel; however, it was after the USA had threatened to withdraw Export-Import Bank backing and export licences for a reactor that the reprocessing agreement was dropped. A small fuel fabrication plant is being built with French assistance.

South Korea signed a safeguards agreement with Australia in 1979, whereby the latter will supply uranium for Korean use. Presumably this uranium will be enriched in the USA. Uranium is also being bought from the USA and Canada. Korea has carried out exploration for uranium in Paraguay during 1978 and 1979. Only small amounts of low grade uranium have been found in Korea itself, despite having begun exploration as long ago as 1955.

It had been hoped at one time that twelve reactors would be operational by 1986 and forty or more by the year 2000. This already begins to seem optimistic.

South Korea is a party to the NPT, having ratified it in 1975 after the USA threatened to withdraw financing for the Kori-2 reactor.

Taiwan
Imported oil has supplied 80% of Taiwan's energy, and 80% of electrical energy has been produced from thermal power

stations (mainly oil-fired). In recent years petroleum and natural gas have been discovered offshore and these should increasingly contribute to indigenous energy supply. Natural gas production is currently of the order of 2000 million cubic metres per year. Coal reserves have been estimated at around 260 million tonnes, and production has been of the order of 3 million tonnes per year. Electric generating capacity is about 8000 MW, three-quarters thermal (including nuclear) and one-quarter hydro.

The Institute of Nuclear Science (at the National Tsing Hua University) was established in 1956 and GE supplied it with a 1 MWth open pool research reactor which became operational in 1961. The Institute of Nuclear Energy Research was founded in 1969 and Canada supplied a 40 MW heavy water natural uranium reactor of the NRX type (i.e. the CIRUS type). This reactor went into operation in 1973. Taiwan now has five research reactors in all, two of them having been built indigenously. A decision to order (from General Electric) two reactors of 600 MWe each was made in the late 1960s, and construction began in 1970. These reactors, Chin-Shan 1 and 2, were commissioned in 1977 and 1979 respectively. Though General Electric supplied the reactors themselves, the generators came from Westinghouse. The cost of both plants together was estimated at around $750 million. Chin-Shan 1 and 2 provide 15% of Taiwan's electricity supply in terms of system capacity. Chin-Shan has produced a larger than expected volume of solid waste, mostly in the form of contaminated clean-up resins. These wastes are mixed with concrete in 55 gallon drums and stored on-site. It is now expected that 3000 drums will be used per year and, already, prepared on-site storage space is running out (Markettos, 1980). By 1981 a total of 14 000 drums had accumulated on site.

Two more reactors of 950 MWe each were ordered from GE in 1972, and two reactors of 900 MWe each were contracted for in 1974 from Westinghouse. The 1900 MWe GE nuclear power station at Kuosheng should near completion in 1982 (fuel loading at Kuosheng-1 was being carried out in January 1981, construction having finished 100 days ahead of schedule). The Westinghouse plants (formerly known as TPC 5 and 6 and now known as Maanshan) are due to be completed by the end of 1985. (The Chinshan and Kuosheng reactors are all within about 20 km of Taipei, while TPC 5 and 6 are sited at the southern end of the island.) Thus, by 1985 it is planned there will be 5000 MWe of nuclear power capacity available, or around one-third of the total expected electric generating capacity (providing about 50% of the total electrical energy production). These three double-reactor plants are just the first phase in the government's nuclear energy programme. The total cost of this first phase is estimated to be $3.6 billion, which seems very cheap since costs elsewhere for a similar programme would be $6 billion or more.

The Taiwan Government announced in June 1980 that it would soon embark on the second phase of the nuclear programme. This is to involve two reactors at Yenliao, to become operational in 1988 and 1989 respectively. They are likely to be 1200 MWe each and their estimated cost has been put at $2.6 billion (again a low price). Bids for these two units were being solicited in 1981.

Taiwan does not appear to have any indigenous uranium supplies (4000 tonnes were bought from South Africa in 1981 for $407 million). A laboratory-scale reprocessing plant was built in the early 1970s, but when the USA discovered this, Taiwan was forced to close it. Taiwan is now in favour of building a regional reprocessing facility on an American island in the Pacific.

Taiwan is a party to the NPT, and the entire programme is subject to IAEA safeguards, even though Taiwan was officially expelled from the IAEA in 1972 when the People's Republic of China was recognised.

Turkey
About 60% of Turkey's commercial energy requirements are supplied by oil, and most of the rest by coal and lignite (Cetelincelik, 1977). In 1970 about 50% of Turkey's oil requirements were supplied domestically, but in 1979 oil production was 3 million tonnes or 20% of requirements. Proven domestic reserves are now expected to run out in fifteen years' time. Proven reserves of natural gas are also small, amounting to 500 billion cubic feet.

Coal production has decreased over recent years and

proven reserves (of 140 million tonnes) are expected to be exhausted by the late 1990s. Lignite reserves, however, stand at 5000 million tonnes, so use of lignite is to be expanded. Production of coal and lignite in million tonnes has been as follows:

	Coal	*Lignite*
1977	8	9
1979	7.3	13.4

Electric generating capacity is around 5000 MW, of which about 3000 is thermal and 2000 is hydro. Total capacity is inadequate to meet demand and severe power cuts are a frequent occurrence. It is planned to increase the hydro capacity available – total hydro potential has been estimated at 16 000 MW.

The Turkish Atomic Energy Commission was founded in 1956 and nuclear research centres were established at Istanbul and Ankara. In 1961 Turkey's first research reactor went critical – a swimming-pool reactor of 1 MW. During 1979 it was announced that Sweden would build a power reactor in Turkey; later, news emerged that agreement had also been reached with the USSR for the supply of another nuclear power station (a 440 MWe PWR type). The agreement with Sweden involved ASEA-ATOM building a 700 MWe BWR reactor at Akkuyu, 35 km from Silifki on the Mediterranean coast. The Turkish Atomic Energy Commision granted a licence for this site in 1976. This Swedish reactor was to be operational by 1987–8, and a second unit was to come on line by 1992. The agreement also allowed the possibility of eight more reactors being built. Each reactor was to cost about $1 billion and 20–30% of this cost would be met by the Turkish government while the rest would be loaned by ASEA-ATOM.

Exploration for radioactive mineral deposits was begun in 1957. Turkey's uranium reserves are now estimated at about 4500 tonnes of fairly low grade, and there may be 300 000 tonnes or more of thorium (Evcimen, 1979), though only a few thousand tonnes are proven. A small uranium mine and mill are in operation, producing one-third of a tonne of uranium per year.

Turkey has signed but not ratified the NPT.

Bibliography

Abraham, P., *et al.* (1971). Atmospheric releases from Tarapur. *Indian Journal of Environmental Health*, July 1971, 171–81.

Abraham, P., Pattnaik, D. and Soman, S. D. (1973). Safety experience in the operation of a BWR station in India. In *Principles and standards of reactor safety*, 459–71, IAEA, Vienna.

Adelman, K. L. and Knight, A. W. (1979). Can South Africa go nuclear? *Orbis*, Fall 1979, 633–47.

Agarwal, A. (1979). Nuclear safety in the Third World. *Nature*, 7 June 1979, 468–70.

Ahmad, A. (1975). India's nuclear age: dubious cost-benefit. *Economic and Political Weekly*, 8 March 1975, 437–44.

Alexander, T. (1975). Our costly losing battle against nuclear proliferation. *Fortune*, December 1975, 143–50.

Allen, B. T. and Melnik, A. (1975). Economics of the power reactor industry. In *Economics of energy*, L. E. Grayson (Ed.), 389–400, Darwin Press, Princeton.

Amalraj, R. V. and Balu, K. (1979). Management of radioactive waste at power reactor sites in India. In *Proceedings of symposium on on-site management of power reactor wastes, Zürich, 26–30 March 1979*. OECD, Paris.

Anderson, R. S. (*c.* 1975). *Building scientific institutions in India: Saha and Bhabha*, Occasional paper no. 11, Centre for Developing-Area Studies, McGill University, Montreal.

Anderson, R. S. and Morrison, B. M. (1974). Power from power: a new scenario emerges for India's scientists. *Science Forum*, December 1974, 10–13.

Atomic Energy Commission (1970). *Atomic energy and space research: a profile for the decade 1970–80*, AEC/Government of India, New Delhi.

Atomic Energy Commission (1961). *Symposium on nuclear power*, AEC, Bombay.

Barber, R. J. (Associates) (1975). *LDC nuclear power prospects 1975–90*. ERDA/NTIS, Springfield, Va.

BARC (1973). *Proceedings of the symposium on nuclear science and engineering* BARC, Bombay.

BARC (1974 to date) *Summary of personnel radiation exposures in BARC and other units of DAE*, BARC, Bombay (annual).

BARC (1977). *Symposium on reactor physics*, BARC, Bombay.

Barretto, P. M. C. (1979). Uranium exploration in developing countries. *IAEA Bulletin*, vol. 21, no. 2/3, 18–28.

Bazin, M. (1980). The politics of power in Brazil. *Nature*, 24 April 1980, 655–7.

146

Becker, K. (1978). Nuclear science and technology in the People's Republic of China (in German). *Atomwirtschaft*, September 1978, 406–7.

Bhabha, H. J. (1956). The role of atomic power in India and its immediate possibilities. In *The economics of nuclear power* J. Gueron *et al.* (Eds), 134–45, Pergamon Press, Oxford.

Bhabha, H. J. (1957). On the economics of atomic power development in India and the Indian atomic energy programme. *Science and Culture*, October 1957, 180–6, and November 1957, 219–25

Bhabha, H. J. (1964). Atomic energy in India. *Science in Parliament*, vol. 2, no. 2, 57–68.

Bhabha, H. J. and Dayal, M. (1962). Some economic aspects of nuclear power in India. *6th World Power Conference, Melbourne*, vol. 10, 4140–62.

Bhargava, G. S. (1978). India's nuclear policy. *India Quarterly* April/June 1978, 131–44.

Bhat, I. S., *et al.* (1974). Population exposure evaluation by environmental measurement and whole-body counting in the environment of nuclear installations (TAPS). In *Population dose evaluation and standards for man and his environment*, 337–46, IAEA, Vienna.

Bhattacharyya, A. (Ed.) (1974). *Proceedings of the seminar on prospects and future of atomic power generation in India*, Institution of Engineers, Calcutta.

Biswas, S. K. (1965). Nuclear development in India. *Nuclear Energy*, March 1965, 87–96.

Bleackley, F. J. (1957). The Canada–India reactor: its design and operating features. *Indian Journal of Power and River Valley Development*, November 1957, 1–19.

Bupp, I. C., Derian, J. C., *et al.* (1975). The economics of nuclear power *Technology Review*, February 1975, 14–25.

Business India (1978). Nuclear power in India. *Business India*, 4–17 September 1978, 20–35.

Business Week (1975). Why atomic power dims today. *Business Week*, 17 November 1975, 98–196.

Business Week (1977). Opposites: GE grows while Westinghouse shrinks. *Business Week*, 31 January 1977, 60–66.

Central Electricity Authority (1977). *Public electricity supply: all-India statistics 1975/76*, Government of India, New Delhi.

Central Water and Power Commission (1961). *Planning for power development in India*, 4th edn, CWPC, New Delhi.

Centre for Monitoring Indian Economy (1979). *Power, coal and oil: review of 1978–9 and prospects*, CMIE, Bombay.

Cetelincelik, M. (1977). Energy resources in Turkey. *Energy International*, September 1977, 27–30 and 41.

Cha, J. H. (1978). Nuclear manpower development in Korea. *Transactions of the American Nuclear Society*, September 1978, 300–9.

Chaudhuri, M. R. (1970). *Power resources of India*, Oxford and IBH Publishing, Calcutta.

Cheema, P. I. (1980). Pakistan's quest for nuclear technology. *Australian Outlook*, August 1980, 188–96.
Chidambaram, R. and Ramanna, R. (1975). Some studies on India's peaceful nuclear explosion experiment. In *Peaceful nuclear explosions*, 421–36, IAEA, Vienna.
Chitale, V. P. (1973). *Foreign technology in India*, Economic and Scientific Research Foundation, New Delhi.
Chitale, V. P. and Roy, M. (1975). *Energy crisis in India*, Economic and Scientific Research Foundation, New Delhi.
Chowdry, K. (Ed.) (1974). *Science policy and national development: Vikram Sarabhai*, Macmillan, New Delhi.
Cilingiroglu, A. (1969). *Manufacture of heavy electrical equipment in developing countries*. IBRD/Johns Hopkins, Baltimore, Md.
Clarke, W. (1980). China's electric power industry (1978). In *Energy in the developing world*, V. Smil and W. E. Knowland (Eds), 145–66, Oxford University Press, London.
Cockroft, J. and Menon, M. G. K. (1967). *Homi Jehangir Bhabha 1909–1966*, Royal Institution, London.
Comitato Nazionale Energia Nucleare (1971). *Tecnica ed Economia della Produzione di Acqua Pesante*, CNEN, Rome.
Commoner, B. (1972). *The closing circle*, Cape, London.
Cook, D. (1978). How Carter's nuclear policy backfired abroad. *Fortune*, 23 October 1978, 124–36.
DAE (1956). *Atomic energy for peaceful purposes in India: proceedings of a conference*, Government of India/DAE, Bombay.
DAE (1970) *Seminar on nuclear power*, BARC for DAE, Bombay.
DAE (1973). *Proceedings of the Indo-Soviet seminar on fast reactors*, DAE, Bombay.
DAE (Engineering Sciences Advisory Committee) (1976). *Proceedings of a symposium on power plant dynamics and control*, DAE, Bombay.
DAE (1977). *Symposium on operating experience of nuclear reactors and power plants*, DAE, Bombay.
DAE (1979). *Symposium on power plant safety and reliability*, DAE, Bombay.
Dasgupta, B. (1971). *The oil industry in India*, Frank Cass, London.
Department of Science and Technology (1977). *Handbook of R and D statistics 1974/5*. Government of India, New Delhi.
Department of Science and Technology (1978). *Handbook of R and D statistics 1976/7*. Government of India, New Delhi.
Desai, B. G. (1978). *Energy policy for India*, Jyoti, Baroda.
Duayer de Souza, M. (1979). *The establishment of nuclear industry in LDCs: the cases of Argentina, Brazil and India*. PhD thesis, University of Manchester.
Ebinger, C. K. (1978). *International politics of nuclear energy*, Sage, London.
Economist (1975). Nuclear trading-go-round. *Economist*, December 6 1975, 74–5.
Egan, J. R. and Arungu-olende, S. (1980). Nuclear power for the Third World. *Technology Review*, May 1980, 46–55.

Eibenschutz, J., Escofet, A. and Bazan, G. (1977). Energy in Mexico. *Transactions of the 10th World Energy Conference, Istanbul,* September 1977.

Energy Survey Committee (1965). *Report,* Government of India, New Delhi.

Estimates Committee (Fourth Lok Sabha) (1970). *129th Report: Department of Atomic Energy,* Lok Sabha, New Delhi.

Estimates Committee (Fifth Lok Sabha) (1973). *29th Report: Department of Atomic Energy,* Lok Sabha, New Delhi.

Evcimen, T. H. (1979). Nuclear safety and nuclear power prospects in Turkey. *Transactions of the American Nuclear Society,* May 1979, 429–31.

Falls, O. B. (1973). A survey of nuclear power in developing countries. *IAEA Bulletin,* vol. 15, no. 5, 27–38.

Fortune (1970). GE's costly ventures into the future. *Fortune,* October 1970, 93.

Fuel Policy Committee (1975). *Report 1974.* Government of India, New Delhi.

Gall, N. (1979). The twilight of nuclear exports. *Ecologist,* October/November 1979 230–5.

Gall, N. (1976). Atoms for Brazil, dangers for all. *Foreign Policy,* vol. 23, Summer 1976, 155–201.

Garrett, P. M. (1978). Current trends in occupational radiation exposures at US commercial power reactors. *Nuclear Engineering International,* April 1978, 51–4.

Gillette, R. (1975a). Nuclear proliferation: India, Germany may accelerate the process. *Science,* 30 May 1975, 911–4.

Gillette, R. (1975b). Uranium enrichment: with help South Africa is progressing. *Science,* 13 June 1975, 1090–2.

Gold, N. L. (1957). *Regional economic development and nuclear power in India,* National Planning Association, Washington, D.C.

Gosling, D. (1973). Hard times for India's nuclear programme. *New Scientist,* 6 September 1973, 564–5.

Gottstein, K. (1977). Nuclear energy for the Third World. *Bulletin of Atomic Scientists,* June 1977, 44–8.

Graham, R. J. and Stevens, J. E. S. (1974). Experience with CANDU reactors outside of Canada: KANUPP and RAPP. *14th Annual International Conference of the Canadian Nuclear Association,* vol. 2, 1–39.

Grenon, M. (1972). L'Energie atomique en Inde. *Revue Française de l'Energie,* February 1972, 246–57.

Grossling, B. (1979). Petroleum experience in developing countries. *Natural Resources Forum (UN),* vol. 3.

Hammond, A. L. (1977). Brazil's nuclear program: Carter's non-proliferation policy backfires. *Science,* 18 February 1977, 657–9.

Hasson, J. A. (1964). Nuclear power in India. *Indian Journal of Economics,* July 1964, 1–29.

Hasson, J. A. (1965). *The economics of nuclear power,* Longmans, London.

Hayes, D. (1977). *Energy for development: Third World options.* Worldwatch Institute, Washington, D.C.

Henderson, P. D. (1975). *India – the energy sector,* Oxford University Press, London.

House, J. (1980). The Third World goes nuclear. *South,* December 1980, 31–5.

Hunt, S. E. (1974). *Fission, fusion and the energy crisis.* Pergamon Press, Oxford.

IAEA (1969 to date). *Operating experience with nuclear power stations in member states,* IAEA, Vienna (annual).

IAEA (1974a). *Market survey for nuclear power in developing countries.* IAEA, Vienna.

IAEA (1974b). *Experience from operating and fuelling nuclear power plants,* IAEA, Vienna.

IAEA (1975). *Nuclear power planning study for Pakistan,* IAEA, Vienna.

IAEA (1977). *International conference on nuclear power and its fuel cycle,* IAEA, Vienna.

IAEA (1978). *International symposium on problems associated with the export of nuclear power plant,* IAEA, Vienna.

IBRD (1968). *Nuclear power for small electricity systems,* World Bank, Washington, D.C.

IBRD (1974). *Nuclear power: its significance for the developing world,* World Bank, Washington, D.C.

IBRD (1979). *A programme to accelerate petroleum production in the developing countries,* World Bank, Washington, D.C.

IBRD (1980). *Energy in developing countries,* World Bank, Washington, D.C.

Indian Association for Radiation Protection (1979). *Safety aspects in the nuclear fuel cycle,* 6th Annual Conference, BARC, Bombay.

Iyengar, B. R. R. and Murthy, S. S. (1972). Towards an all-India power grid. *Journal of Central Board of Irrigation and Power (India),* April 1972, 155–68.

Jain, J. P. (1974). *Nuclear India* (2 volumes), Radiant Publishers, New Delhi.

Johnson, L. A. (1978). Occupational radiation exposures at LWRs 1969 to 1976. *Nuclear Safety,* November/December 1978, 760–5.

Kamath, P. R. (1973). Backfitting the site to changing radiation environment. In *Principles and standards of reactor safety,* 245–52, IAEA, Vienna.

Kamath, P. R., Bhat, I. S., Ganguly, A. K. (1971). Environmental behaviour of discharged radioactive effluents at TAPS. In *Environmental aspects of nuclear power stations,* IAEA, Vienna.

Kapur, A. (1976). *India's nuclear option: atomic diplomacy and decision making.* Praeger, New York.

Kapur, A. (1980). A nuclearizing Pakistan: some hypotheses. *Asian Survey,* May 1980, 495–516.

Kashfi, M. S. (1980). The energy alternative for Iran's future: nuclear power. *Impact of Science on Society,* vol. 30, no. 3, 217–22.

Bibliography 151

Kashkari, C. (1975). *Energy: resources, demand and conservation with special reference to India*, Tata–McGraw-Hill, New Delhi.

Kaul, R. (1974). *India's nuclear spin-off*, Chanakya, Allahabad.

Khalilzad, Z. (1980). Pakistan and the Bomb. *Bulletin of Atomic Scientists*, January 1980, 11–16.

Khera, S. S. (1979). *Oil: rich man, poor man*, National Publishing House, New Delhi.

Krugmann, H. (1981). The German–Brazilian nuclear deal. *Bulletin of Atomic Scientists*, February 1981, 32–6.

Kulkarni, R. P. and Sarma, V. (1969). *Homi Bhabha: father of nuclear science in India*. Popular Prakashan, Bombay.

Kumaramangalam, S. M. (1973). *Coal industry in India*. Oxford and IBH Publishing, Calcutta.

Lamarsh, J. R. (1980). China's nuclear power program. *Bulletin of Atomic Scientists*, May 1980, 28–31.

Lee, I. H., *et al.* (1977). Fuel cost analysis of CANDU-PHWR Wolsung Unit 1. *Hanguk Wonjaryok Hakhoe (Journal of Korean Nuclear Society)*, September 1977, 151–163.

Lefever, E. W. (1979). *Nuclear arms in the Third World*, Brookings Institution, Washington, D.C.

Lesurf, J. E. (1977). Control of radiation exposures at CANDU nuclear power stations. *Journal of British Nuclear Energy Society*, January 1977, 53–61.

Leung, C. T. (1980). China and Hong Kong: feasibility of a joint nuclear project. *Energy World*, July 1980, 7–10.

Lloyds Bank (1981). *Country report: Iraq*, Lloyds Bank, London.

Lopes, J. L. (1978). Atoms in the developing nations. *Bulletin of Atomic Scientists*, April 1978, 31–4.

Lovins, A. B. (1975). *Nuclear power: technical bases for ethical concern*, 2nd edn, Friends of the Earth, London.

Lowrence, W. W. (1976). Nuclear futures for sale: to Brazil from West Germany, 1975. *International Security*, vol. 1, no. 2, 163.

Lumb, P. B. (1976). The Canadian heavy water industry. *Journal of British Nuclear Energy Society*, January 1976, 35–46.

McIntyre, H. C. (1975). Natural uranium heavy water reactors – CANDU. *Scientific American*, October 1975, 17–27.

McNeil, M. (1979). *Brazil's U/Th deposits: geology, reserves, potential*, Miller Freeman, San Francisco.

Mahatme, D. B. (1975). Heavy water for self-reliant growth. *Commerce*, 10 May 1975, 691–3.

Markettos, N. D. (1980). Ambitious nuclear programme planned for Taiwan. *Nuclear Engineering International*, March 1980, 13–14.

Masters, R. (1976). AECL–industrial profile. *Nuclear Engineering International*, June 1976, 49–57.

Meckoni, V. N. and Balakrishnan, M. R. (1974). The potential of nuclear fuel in meeting the energy needs of India. *Physics News*, vol. 5, no. 3, 95–99.

Mehta, B. (1974). *India and the world oil crisis*, Sterling Publishers, New Delhi.

Miccolis, J. M. (1978). Alternative energy technologies in Brazil. In *Renewable energy resources and rural applications in the developing world*, N. L. Brown (Ed.), Westview Press, Boulder, Colo.

Miller, S. (1976). *The economics of nuclear and coal power*, Praeger, New York.

Ministry of Energy (1979). *Report 1978/79*, Government of India, Ministry of Energy (Department of Power), New Delhi.

Mirchandani, G. G. (1968). *India's nuclear dilemma*, Popular Book Service, New Delhi.

Morehouse, W. (1971). *Science in India*, Popular Prakashan, Bombay.

Mossavar-Rahmani, B. (1980). Iran's nuclear power programme revisited. *Energy Policy*, September 1980. 189–202.

Nanda, B. R. (Ed.) (1977). *Science and technology in India*. Vikas, New Delhi.

National Committee on Science and Technology (1974). *Report of the fuel and power sector*, Government of India, NCST, New Delhi.

National Council of Applied Economic Research (1962). *Demand for energy in southern India*. NCAER, New Delhi.

National Council of Applied Economic Research (1966). *Demand for energy in India*, NCAER, New Delhi.

National Productivity Council and Central Water and Power Commission (1973). *Energy system economics*, National Productivity Council, New Delhi.

Nau, H. R., et al. (1976). *Technology transfer and US foreign policy*, Praeger, New York.

Nehrt, L. C. (1966). *International marketing of nuclear power plants*, Indiana University Press, Boise, Ind.

Netschert, B. C. and Schurr, S. H. (1957). *Atomic energy applications with reference to underdeveloped countries*, Resources for the Future/Johns Hopkins Press, Baltimore, Md.

Newby-Fraser, A. R. (1979). *Chain reaction: 20 years of nuclear research and development in South Africa*, AEB, Pretoria.

OECD/NEA and IAEA (1979). *Uranium: resources, production and demand*, OECD/NEA and IAEA.

Overholt, W. H. (Ed.) (1977). *Asia's Nuclear Future*, Westview Press, Boulder, Colo.

Pachauri, R. K. (1977). *Energy and economic development in India*, Praeger, New York.

Pachauri, R. K. (Ed.). *International energy studies*, Wiley Eastern Ltd, New Delhi.

Parikh, S. S. (Ed.) (1977). *Coal after nationalisation*, Coal Consumer's Association of India, Calcutta.

Parrott, M. (1980). Korea: an expanding economy trying to gain energy independence. *Nuclear Engineering International*, February 1980, 13–14.

Pathak, K. K. (1980). *Nuclear policy of India: a Third World Perspective*, Gitanjali Prakashan, New Delhi.

Patterson, W. C. (1980). *Nuclear power*, revised edn, Penguin, Harmondsworth.

Perez-Lopez, J. F. (1979). The Cuban nuclear power programme. *Cuban Studies*, January 1979, 1–42.
Perry, W. and Kern, S. (1978). The Brazilian nuclear programme in a foreign policy context. *Comparative Strategy*, vol. 1, nos 1–2, 53–70.
Planning Commission (1978). *Draft five year plan 1978–83*, Government of India, New Delhi.
Poulose, T. T. (1979). India and the nuclear safeguards controversy. *India Quarterly*, April–June 1979 153–62.
Power, P. F. (1979). The Indo-American nuclear controversy. *Asian Survey*, June 1979, 574–96.
Power Economy Committee (1971). *Report*, Government of India, New Delhi.
Rae, H. K. (1978). Selecting heavy water processes. In *Separation of hydrogen isotopes*, H. K. Rae (Ed.), 1–26, American Chemical Society, Washington, D.C.
Rahman, A. (1974). *Science technology and economic development*, National Publishing, New Delhi.
Rahman, A. and Sharma, K. D. (Eds) (1974). *Science Policy Studies*, Somaiya, Bombay.
Ramanna, R. (1974). New possibilities of the peaceful uses of atomic energy. *Chemical Engineering World 1974*, 37–42, 45–6 and 49–54.
Ramanna, R. (1975). Development of nuclear energy in India. *Electrical India*, 31 May 1975, 13–18B.
Ramanna, R. (1976). Safety of nuclear installations. *Electrical India*, 15 April 1976, 17–27.
Rao, N. K. (1975). Nuclear fuel fabrication in India. *Transactions of the American Nuclear Society*, 21 April 1975, 407–10.
Rao, R. R. (1974) India's nuclear progress – a balance sheet. *India Quarterly*, October–December 1974, 239–53.
Redick, J. R. (1972). *Military potential of Latin American nuclear energy programmes*, Sage, New York.
Redick, J. R. (1978). Regional restraint: US nuclear policy and Latin America. *Orbis*, Spring 1978, 161–200.
Redick, J. R. (1981). The Tlatelolco regime and nonproliferation in Latin America. *International Organisation*, vol. 35, no. 1, 103–34.
Richardson, M. (1978). Manila's disillusion with the atom. *Far Eastern Economic Review*, 23 June 1978, 94–9.
Riley, P. J. and Darracott, J. M. (1980). Construction of Kori nuclear power station No. 1. *Nuclear Energy*, October 1980, 369–79.
Robertson, J. A. L. (1978). The CANDU reactor system: an appropriate technology. *Science*, vol 199, no 4329, 657–64.
Rosen, M. (1977). The critical issue of nuclear power plant safety in developing countries. *IAEA Bulletin*, April 1977, 12–21.
Rosen, M. (1979). Nuclear power in developing countries: the transfer of regulatory capability. *IAEA Bulletin*, vol. 21, no. 2/3, 2–12.
Rosenbaum, H. J. (1976). Brazil's nuclear aspirations. In *Nuclear proliferation and the near-nuclear countries*, O. Marwah and A. Schulz (Eds), 255–77. Ballinger, Cambridge, Mass.
Rosenstein-Rodan, P. N. (1964). Contribution of atomic energy to a

154   *Nuclear Power in India*

power programme in India. In *Pricing and fiscal policies*, P. N. Rosenstein-Rodan (Ed.), 165–74, Allen and Unwin, London.

Rowe, J. W. (1969). Nuclear energy policy in Brazil. In *The social reality of scientific myth*, K. H. Silvert (Ed.), American Universities Field Staff.

Roy, R. (1980). *Family and community biogas plants in rural India*. Open University Working Paper, Milton Keynes.

Royal Commission on Electric Power Planning (Chairman: Arthur Porter) (1978). *Interim report on nuclear power in Ontario*, RCEPP, Toronto.

Sabato, J. A. (1973). Atomic energy in Argentina: a case history. *World Development*, August 1973, 23–38.

Sabato, J. A. and Frydman, R. J. (1977). Latin America goes nuclear. *Atlas World Press Review*, June 1977, 26–7.

Sabato, J. A. and Ramesh, J. (1980). Atoms for the Third World. *Bulletin of Atomic Scientists*, March 1980, 36–43.

Sarabhai, V. A. (1969). Why nuclear power for India? *Science Today* (*New Delhi*). September 1969, 32–7.

Sardar, Z. (1981). Why the Third World needs nuclear power. *New Scientist*, 12 February 1981, 402–4.

Sarma, M. S. R., *et al.* (1975). Maintenance experience at Rajasthan atomic power station. In *Reliability of nuclear power plants*, 689–701, IAEA, Vienna.

Schiffer, M. (Ed.) (1976). *Fourth Indo-German seminar on operation of nuclear power plants*, Kernforschungsanlage, Jeulich.

Schulten, R. (1980). Nuclear research and nuclear energy in China (in German). *Atomwirtschaft*, April 1980, 197–8.

Seshagiri, N. (1975). *The bomb: fallout of India's nuclear explosion*. Vikas, New Delhi.

Seth, S. P. (1975). India's atomic profile, *Pacific Community*, January 1975, 272–82.

Sethna, H. N. (1972a). *Atomic energy*, Government of India, Ministry of Information, New Delhi.

Sethna, H. N. (1972b). India: past achievements and future promises. *IAEA Bulletin*, vol. 14, no. 6, 36–44.

Sethna, H. N. (1977). Fast breeder reactors for nuclear power. *Urja*, December 1977, 191–7.

Sethna, H. N. and Srinivasan, M. R. (1977). Constraints in implementation of India's nuclear power programme. *Urja*, vol. 2, no. 3, 88–92 and 97.

Shah, J. C. (1978). Role of nuclear power in the energy problem. *Electrical India*, 30 November 1978, 25–30.

Simnad, M. (1981). Progress in the developing countries. *Nuclear Engineering International*, February 1981, 29–34.

Singh, J. (1973). Planning and organisation for atomic energy research. In *Management of scientific research*, J. Singh (Ed.), Popular Prakashan, Bombay.

SIPRI (1978). *Arms control: a survey and appraisal of multilateral agreements*, Taylor and Francis, London.

Smil, V. (1976). *China's energy: achievements, problems, prospects.* Praeger, New York.

Smil, V. and Knowland, W. E. (1980). *Energy in the developing world*, Oxford University Press, London.

Smith, D. (1980). *South Africa's nuclear capability*, World Campaign against Military and Nuclear Collaboration with South Africa, Oslo.

Srinivasan, M. R. (1976). Role of atomic power in electrical power production in India. *Journal of Physics Education (India)*, vol. 3, no. 4, 1–6.

Subrahmanyam, K. (1975). India's nuclear policy. In *Nuclear proliferation and the near-nuclear countries*, O. Marwah and A. Schulz (Eds), 125–48, Ballinger, Cambridge, Mass.

Surrey, J. and Thomas, S. (1980). Worldwide nuclear plant performance. *Futures*, February 1980, 3–17.

Swayambu, S. (1972). *Power development*, Government of India, New Delhi.

Sweet, W. (1978). US–India safeguards dispute. *Bulletin of Atomic Scientists*, June 1978, 50–2.

Tahir-Kheli, S. (1978). Pakistan's nuclear option and US policy, *Orbis*, Summer, 1978, 357–74.

Thomas, K. T. (1976). Energy options and the role of nuclear energy in Asian countries. In *Facing up to nuclear power*, J. Francis and P. Abrecht (Eds), St Andrews Press, Edinburgh.

Thomas, K. T., *et al.* (1970). *The nuclear powered agro-industrial complex: report of a working group*, BARC, Bombay.

Tomar, R. (1980). The Indian nuclear power programme – myths and mirages. *Asian Survey*, vol. 20, no. 5, 517–31.

Tyner, W. E. (1978). *Energy resources and economic development in India*, Martinus Nijhoff, Leiden and Boston.

UN (1955). *First international conference on the peaceful uses of atomic energy*, UN, New York.

UN (1958). *Second international conference on the peaceful uses of atomic energy*, UN, New York.

UN (1964). *Third International conference on the peaceful uses of atomic energy*, UN, New York.

UN (1972). *Fourth international conference on the peaceful uses of atomic energy*, UN, New York.

UNCTAD (1980). *Energy supplies for developing countries*; UN, New York.

UNESCO (1972). *National science policy and organisation of scientific research in India*, UNESCO, Paris.

US Bureau of the Census (1980). *Statistical abstract of the US: 1980*, US Government Printing Office, Washington, D.C.

Venkataraman, K. I. (1972). *Power development in India: the financial aspects*, Halsted, New York.

Wadia, D. N. (1957). India and the atomic age – her natural resources for nuclear power production. *Science and Culture*, December 1957, 264–70.

Wagle, D. M. and Rao, N. V. (1978). *Power sector in India*, Popular Prakashan, Bombay.

Wall Street Journal (1976). GE studies nuclear – marketing changes to limit liability, raise potential profit. *Wall Street Journal*, 18 November 1976, 2.

Wasserman, H. and Wainer, A. (1976). Nuclear power to the Pacific. *Environment*, November 1976, 18–20 and 25.

Williams, S. L. (1969). *The US, India and the Bomb*, Johns Hopkins, Baltimore, Md.

Willrich, M. (1971a). *Global politics of nuclear energy*, Praeger, New York.

Willrich, M. (Ed.) (1971b) *Civil nuclear power and international security*, Praeger, New York.

Wilson, E. M. (1977). Tidal power in India. *Indian Journal of Power and River Valley Development*, May 1977, 141.

Wonder, E. F. (1977a). Nuclear commerce and nuclear proliferation: Germany and Brazil 1975. *Orbis*, vol. 21, no. 2, 277–307.

Wonder, E. F. (1977b). *Nuclear fuel and American foreign policy: multilateralisation for uranium enrichment*, Westview Press, Boulder, Colo.

World Energy Conference (1974). *Survey of energy resources 1974.* WEC, London.

Zaheer, S. H. (1974). India and the energy crisis. *Chemical Age of India*, September 1974. 609–14.

Zaheer, S. H. (1975). Thoughts on meeting the energy crisis. *Chemical Age of India*, June 1975, 437–40.

Index